Lecture Notes in Control and Information Sciences 392

Editors: M. Thoma, F. Allgöwer, M. Morari

Bijnan Bandyopadhyay, Fulwani Deepak,
and Kyung-Soo Kim

Sliding Mode Control Using Novel Sliding Surfaces

Springer

Authors

Prof. Bijnan Bandyopadhyay

114 B, IDP in Systems and
Control Engineering
CRNTS Building
IIT Bombay
Powai, Mumbai 400076
India
E-mail: bijnan@ee.iitb.ac.in

Prof. Kyung-Soo Kim

Department of Mechanical Engineering
KAIST
335 Gwahang-no
Yuseaong-gu, Daejeon 305-701
Korea
E-mail: kyungsookim@kaist.ac.kr

Fulwani Deepak

101C, IDP in Systems and
Control Engineering
CRNTS Building
IIT Bombay
Powai, Mumbai 400076
India
E-mail: fulwani@gmail.com

ISBN 978-3-642-03447-3 e-ISBN 978-3-642-03448-0

DOI 10.1007/978-3-642-03448-0

Lecture Notes in Control and Information Sciences ISSN 0170-8643

Library of Congress Control Number: 2009933690

Typeset & Cover Design: Scientific Publishing Services Pvt. Ltd., Chennai, India.

Printed in acid-free paper

5 4 3 2 1 0

springer.com

Thou hast made me endless, such is thy pleasure. This frail
vessel thou emptiest again and again, and fillest it ever with fresh life.
This little flute of a reed thou hast carried over hills and dales,
and hast breathed through it melodies eternally new.

Rabindranath Tagore

To my father Nitya Niranjan Bandyopadhyay

Bijnan Bandyopadhyay

To the memory of my father Maganlal Fulwani

Deepak Fulwani

To my parents and family

Kyung-Soo Kim

Preface

After a survey paper by Utkin in the late 1970s, sliding mode control methodologies emerged as an effective tool to tackle uncertainty and disturbances which are inevitable in most of the practical systems. Sliding mode control is a particular class of variable structure control which was introduced by Emel'yanov and his colleagues. The design paradigms of sliding mode control has now become a mature design technique for the design of robust controller of uncertain system. In sliding mode technique, the state trajectory of the system is constrained on a chosen manifold (or within some neighborhood thereof) by an appropriate control action. This manifold is also called a switching surface or a sliding surface. During sliding mode, system dynamics is governed by the chosen manifold which results in a well celebrated invariance property towards certain classes of disturbance and model mismatches.

The purpose of this monograph is to give a different dimension to sliding surface design to achieve high performance of the system. Design of the switching surface is vital because the closed loop dynamics is governed by the parameters of the sliding surface. Therefore sliding surface should be designed to meet the closed loop specifications. Many systems demand high performance with robustness. To address this issue of achieving high performance with robustness, we propose nonlinear surfaces for different classes of systems. The nonlinear surface is designed such that it changes the system's closed-loop damping ratio from its initial low value to a final high value. Initially, the system is lightly damped resulting in a quick response and as the system output approaches the setpoint, the system is made overdamped to avoid overshoot. The technique of using a nonlinear surface to improve performance is discussed in four chapters after introducing basic notions in the introductory chapter. To address performance objectives in sliding mode, a generalized framework for sliding surface design is presented in last two chapters for continuous and discrete time systems, based on the full order Lyapunov matrices which constitute Lyapunov- or Riccati-type inequalities. The same method ensures performance in the face of parametric uncertainty which does not satisfy the matching condition. It also enables to optimize the sliding motion by applying the guaranteed cost control idea.

The authors would like to express their deep sense of gratitude to their parents and teachers who have made them capable enough to write this book. The authors wish to thank many individuals who had helped them directly or indirectly in completing this monograph. In particular authors wish to express their thanks to several fellow researchers of Syscon, IIT Bombay and Mechanical Engineering Department, KAIST. Finally authors wish to acknowledge the support, patience and love of their wives and children during the preparation of this monograph.

June 2009 Bijnan Bandyopadhyay
IIT Bombay, India Deepak Fulwani
KAIST, Daejeon, Republic of Korea Kyung-Soo Kim

Contents

Acronyms

Abbreviations

CNF	Composite Nonlinear Feedback
DSM	Discrete-time Sliding Mode
DSMC	Discrete-time Sliding Mode Control
ISM	Integral Sliding Mode
ISMC	Integral Sliding Mode Control
LQR	Linear Quadratic Regulator
LTI	Linear Time Invariant
LMI	Linear Matrix Inequality
ISMC	Integral Sliding Mode Control
QSM	Quasi Sliding Mode
QSMB	Quasi Sliding Mode Band
VSCS	Variable Structure Control System

List of Frequently Used Symbols

A System matrix of continuous time LTI system
B Input matrix of Continuous time LTI system
C_1, C Output matrix
x State vector
u Control input
z State vector in regular form
c^T Sliding surface parameters
ρ, d System uncertainty/disturbances
c_1, c_2 Sliding surface parameters
A_{11}, A_{12}, A_{21}, A_{22} System matrices in regular form
Q, K Positive definite diagonal matrix
τ, Δ Sampling time
η Positive constant

$\mathbb{R}$ The field of real numbers
$\mathbb{R}^n$ The real vector space of dimension-n
Φ System matrix of discrete-time LTI system
Γ Input matrix of discrete-time LTI system
Γ_2 Nonsingular square matrix
T_r Orthogonal transformation matrix
d Disturbances/uncertainty
s switching function
Ψ Nonlinear function
ζ_1, ζ_2 System damping ratios
P, W Positive definite matrices
β Positive constant/matrix
$\bar{k}$ Positive constant
V, V_1 Positive definite function
e Error between actual and desired trajectory
z_d Desired trajectory
A_δ System matrix represented with delta operator
B_δ Input matrix represented with delta operator
h An integer accounts for delay in input
D Disturbance matrix
$\hat{x}$ Predicted state vector
n Number of states of an LTI system model
m Number of input of an LTI system model

List of Figures

Chapter 1
Introduction

1.1 Sliding Mode Control

Beginning in the late 1970s and continuing today, the sliding mode control has received plenty of attention due to its insensitivity to disturbances and parameter variations. The well known *sliding mode control* is a particular type of Variable Structure Control System (VSCS). Recently many successful practical applications of sliding mode control (SMC) have established the importance of sliding mode theory which has mainly been developed in the last three decades. This fact is also witnessed by many special issues of learned journals focusing on sliding mode control [13, 15]. The research in this field was initiated by Emel'yanov and his colleagues [38, 39], and the design paradigm now forms a mature and an established approach for robust control and estimation. The idea of sliding mode control (SMC) was not known to the control community at large until an article published by Utkin [96] and a book by Itkis [52].

SMC design can be divided into two subparts viz. (1) the design of a stable surface and (2) the design of a control law to force the system states onto the chosen surface in finite time. The design of the surface should address all constraints and required specifications therefore it should be designed optimally to meet all requirements. To eliminate the non-robust reaching phase, an *integral sliding mode* was proposed in [100, 19] which naturally allows SMC to be combined with other techniques; this is very important from a practical point of view. In [11, 27], time varying sliding surface design is proposed to eliminate the reaching phase. In [28] a time varying sliding surface is designed for the elimination of the reaching phase and to avoid actuator saturation. The effectiveness of SMC in the robust control of linear uncertain systems prompted the research of sliding mode control in other types of systems. Thus, a few researchers worked on the sliding mode control of nonlinear systems and time delay systems [104, 106]. To relax the need for measuring the entire state vector, an output feedback based sliding mode concept is also proposed in [34, 35, 33] which widens the scope of sliding mode control.

This chapter presents a brief introduction to the basic notions of continuous and discrete-time sliding mode control which will be used in subsequent chapters. We

B. Bandyopadhyay et al.: Sliding Mode Control Using Novel Sliding Surfaces, LNCIS 392, pp. 1–15.
springerlink.com

do not aim for an in-depth and complete presentation of existing results on sliding mode, but we attempt to introduce basic notions in a simplified approach by beginning with a motivational example of a first order uncertain system.

Consider a first order uncertain system modeled as

$$\dot{x}(t) = ax(t) + bu(t) + \rho(x,t), \tag{1.1}$$

where $x(t) \in \mathbb{R}$, $u(t) \in \mathbb{R}$, and a, b are known nonzero constants. The term $\rho(x,t) \in \mathbb{R}$ accounts for the uncertainty and only the bounds of this uncertain term are known. The control objective is to stabilize $x(t)$ when only the bounds of uncertainty are known. To stabilize the uncertain system in (1.1), if initial value of $x(t)$ is positive then $\dot{x}(t)$ should be negative and vice versa. Therefore, depending on the sign of $x(t)$, control law should be *altered* to ensure stabilization of $x(t)$. Consider a control law:

$$u(t) = -b^{-1}(ax(t) + Qsign(x)), \tag{1.2}$$

where $sign(.)$ represents the signum function, and $Q > 0$ is chosen such that

$$|\rho(x,t)| \leq Q. \tag{1.3}$$

With control law (1.2), system (1.1) becomes

$$\dot{x}(t) = -Qsign(x(t)) + \rho(x,t). \tag{1.4}$$

To analyze the above closed loop system, consider the case when initial condition $x(0) > 0$. Due to condition (1.3), it follows that $\dot{x}(t) < 0$. Therefore, $x(t)$ is decreasing and moving towards $x(t) = 0$. When initial condition $x(0) < 0$, then using condition (1.3), it implies that $\dot{x} > 0$. Therefore $x(t) > 0$ and approaches $x(t) = 0$. So in this case also $x(t)$ is moving towards the line $x(t) = 0$. Thus, irrespective of the initial condition, with control law (1.2), the system state $x(t)$ is forced towards $x(t) = 0$. The control law ensures a minimum rate of decrease (or increase) of $x(t)$ therefore $x(t)$ reaches in finite time. At $x(t) = 0$, the discontinuous part of the control law is not defined. However, the moment the trajectory crosses $x(t) = 0$ from either direction, again it is forced back on $x(t) = 0$. This situation is explained pictorially in Fig. 1.1 which shows the state trajectories in the vicinity of $s(x,t) = 0$. Because the control law is discontinuous about $x(t) = 0$ it demands switching at very high frequency. If this switching occurs at a very high frequency (more precisely at infinite frequency) then $x(t) = 0$ can be consistently maintained with this discontinuous control law. The initial phase when the trajectory is forced towards $x(t) = 0$ is called the *reaching phase* and the phase when $x(t) = 0$ is called the *sliding phase* or *sliding mode*. During the sliding phase, with this discontinuous control law, $x(t) = 0$ is maintained even in presence of consistent perturbations. Therefore system motion is insensitive to perturbations. It should be noted that during the reaching phase, perturbations can affect the system performance. Now to verify this analysis, we simulate a numerical example.

For the system in (1.1) let us assume the following parameters:
$a = 1,\ b = 1, \rho(x,t) = 0.2sin(5t)x(t) + 0.3sin(20t)$

By substituting the above parameters, (1.1) becomes

$$\dot{x}(t) = x(t) + u(t) + 0.2sin(5t)x(t) + 0.3sin(20t).$$

With $Q = 0.3\|x\| + 0.4$, condition (1.3) is satisfied. Let us assume initial condition $x(0) = 2$. Using these parameters control law (1.2) is implemented. Fig. 1.2 shows the evolution of state $x(t)$ with time. It can be seen from this figure that in spite of persistent disturbances, $x(t)$ reaches to zero in a finite time and thereafter it remains zero. In this simulation, the control law can switch at a very high frequency. When the control law switching frequency is constrained, it is not possible to maintain ideal sliding motion; the control law $u(t)$ and surface $s(x,t)$ both chatter as seen in Fig. 1.4 and Fig. 1.5 respectively. Fig. 1.3 shows the plot of input $u(t)$ with time which changes between two values from $t = 3.1sec.$ onwards. This happens because the trajectory can not remain on the line $x(t) = 0$ due to the presence of persistence disturbances and thus, as explained earlier, the discontinuous control pushes the state trajectory from either side onto $x(t) = 0$. This process continues and the control law switches between two values. This phenomena is known as *chattering* which needs to be addressed appropriately. The line $x(t) = 0$ across which the control law switches is called *sliding surface* or *sliding hyperplane*. Usually it is denoted by the symbol s and defined as $s(x,t) \equiv c^T x$. For the first order system which is considered above, $c^T = 1$ and $s(x,t) = x(t)$.

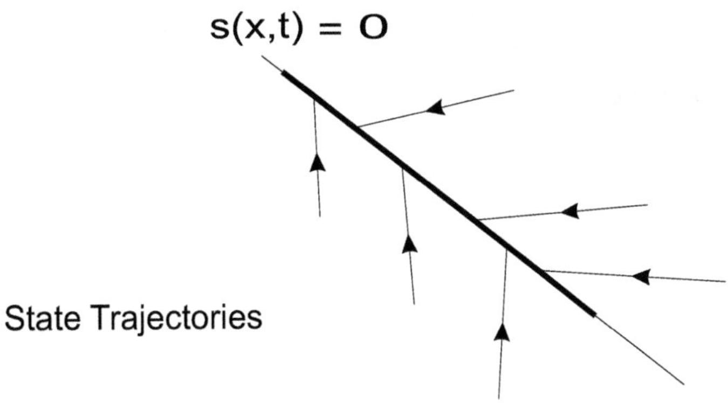

Fig. 1.1 State trajectory in the vicinity of $s(x,t) = 0$

In the above analysis we quantitatively presented the behavior of (1.4). The closed loop system in (1.4) contains a discontinuous function which is non-Lipschitz; therefore classical techniques can not be used for analysis. One way to solve a differential equation with a discontinuous right hand side is proposed by Filippov [41]. To understand this notion in brief, let us rewrite (1.4) as

$$\dot{x} = f(x,u,t) \tag{1.5}$$

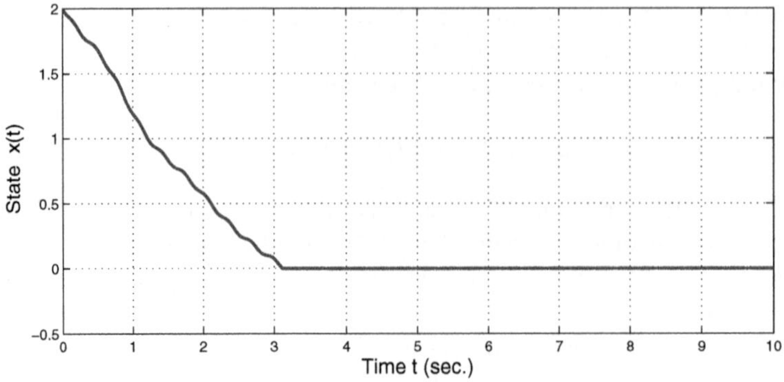

Fig. 1.2 Evolution of state $x(t)$ with time

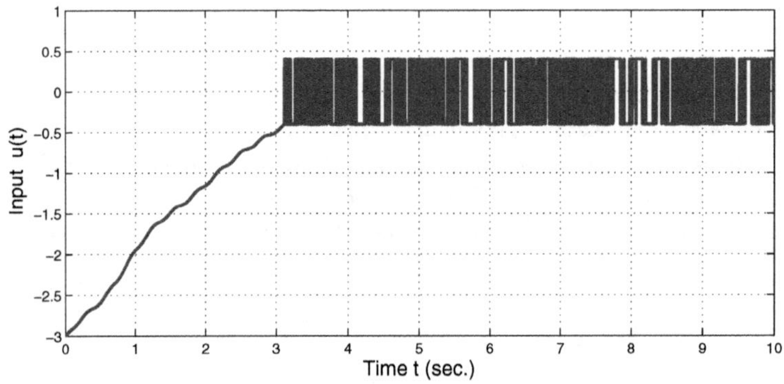

Fig. 1.3 Plot of control input $u(t)$ with time

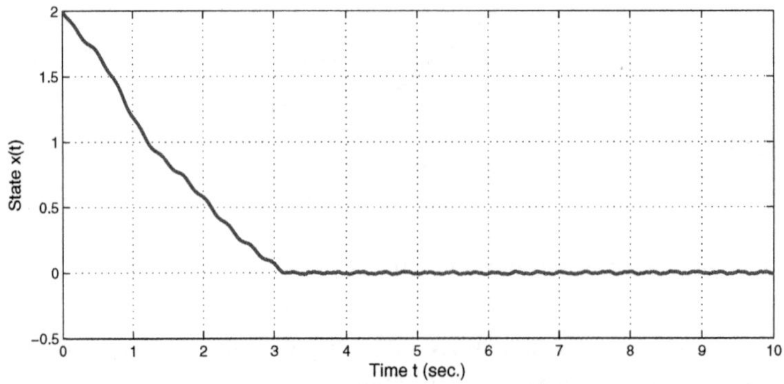

Fig. 1.4 Evolution of state/surface with low switching frequency

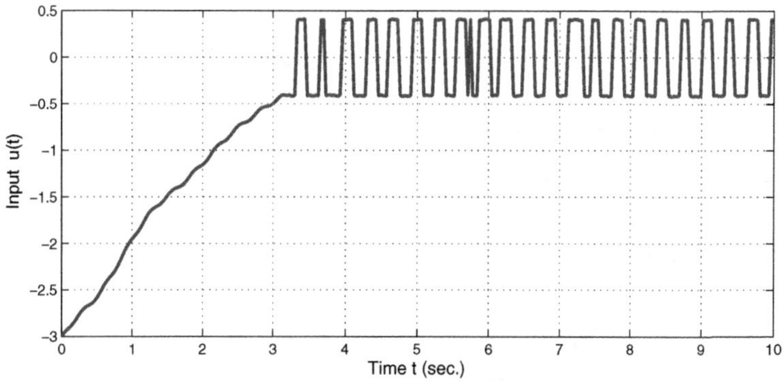

Fig. 1.5 Plot of control input $u(t)$ with a low switching frequency

where $f(x,u,t) = -Qsign(x(t)) + \rho(x,t)$. Suppose at t_1 second the trajectory intercepts sliding surface and sliding mode occurs. During sliding mode, $s(x,t) = x(t) = 0$. For any time $t_2 \geq t_1$ let us denote as f_- and f_+ the value of the function $f(x,u,t)$ when it approaches the discontinuity surface $x(t) = 0$ from either direction as shown in Fig.1.6. Filippov's approach uses the *average* value of f_- and f_+ to obtain the solution. The solution is obtained as

$$\dot{x} = (1-\nu)f_- + \nu f_+ \tag{1.6}$$

where the scalar $0 \leq \nu \leq 1$ is selected such that

$$f_{av} \equiv (1-\nu)f_- + \nu f_+$$

remains tangential to sliding surface $s(x,t) = x(t)$. This approach can be used to analyse motion during sliding mode because classical theory of differential equations is not applicable.

A detailed analysis is presented in [97]. Recent developments in this domain are summarized in a comprehensive tutorial on discontinuous dynamical system by Jorge Cortés [29].

The other approach to analyze motion during sliding mode, which is widely accepted by the community, is *equivalent control.* In this approach, the discontinuous control (which brings sliding mode $s(x,t) = 0$) is replaced by a linear control. The linear control is called *equivalent control* and is obtained by setting $\dot{s}(x,t) = 0$ for the nominal system. For the system in (1.4) equivalent control becomes

$$u_{eq} = -b^{-1}ax(t). \tag{1.7}$$

To analyze system motion during sliding mode, the discontinuous control can be replaced by the equivalent control. It should be noted that in practice due to finite switching frequency, switch imperfections and presence of unmodelled dynamics, ideal sliding mode is not possible. In practice, the state trajectory confines within a boundary layer around $s(x,t) = 0$. However, the equivalent control approach can be

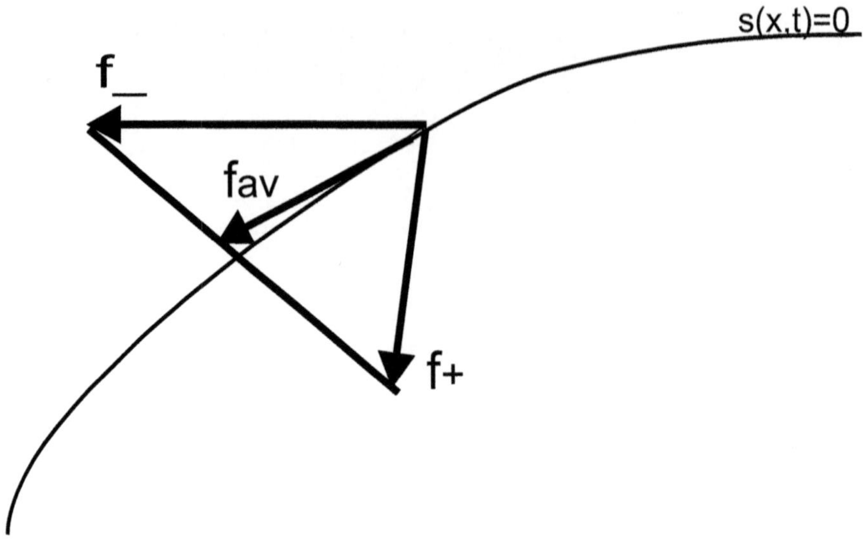

Fig. 1.6 Filippov's method to analyze sliding mode

used to analyze problems which can not be analyzed by Filippov's approach [99]. Utkin *et al.* [99] have presented a detailed analysis of this approach. Physically, equivalent control is the average of discontinuous control which occurs at high frequency to ensure $s(x,t) = 0$.

Formally, a possible definition of sliding mode is as follows:

Definition 1.1.1. *(**Sliding Mode**) Sliding motion or sliding mode may be defined as the evolution of the state trajectory of a system confined to a specified non-trivial sub-manifold of the state space with stable dynamics.*

In what follows, a method to design the sliding surface for general uncertain linear system, and a conceptual framework for other possible methods to design sliding surface, are presented.

1.1.1 Regular Form

Consider an uncertain linear system described as

$$\dot{x}(t) = Ax(t) + B(u(t) + \rho(x,t)) \tag{1.8}$$

$$y(t) = C_1 x(t) \tag{1.9}$$

Where $x(t) \in \mathbb{R}^n$, $u(t) \in \mathbb{R}$, $y(t) \in \mathbb{R}$ are respectively the state, input, and output of the system. A, B, C_1 are matrices of appropriate dimensions. It is worth noting that uncertainty $\rho(x,t)$ lies in the range space of matrix B and therefore it is a matched uncertainty. For illustrative purposes only a matched perturbation is considered.

Appropriate treatment of unmatched perturbation will be given in the subsequent chapters. We make the following assumptions.
Assumption A_1 Pair (A,B) is stabilizable.
Assumption A_2 Matrix B is of full rank i.e $rank(B) = m$.
By assumption of A_2 there exists a transformation matrix T_r such that

$$\dot{z}_1 = A_{11}z_1 + A_{12}z_2, \tag{1.10}$$
$$\dot{z}_2 = A_{21}z_1 + A_{22}z_2 + B_2u + \rho_2(z,t), \tag{1.11}$$
$$y = Cz, \tag{1.12}$$

where $z_1 \in \mathbb{R}^{n-m}$, $z_2 \in \mathbb{R}^m$, $C = C_1(T_r)^{-1}$, $\rho_2(z,t) = T_r\rho(x,t)$, B_2 is an m by m matrix and all other matrices are of appropriate dimensions. The sliding surface in z coordinate can be designed as

$$s = \begin{bmatrix} c_1 & c_2 \end{bmatrix} \begin{bmatrix} z_1 \\ z_2 \end{bmatrix}. \tag{1.13}$$

Let us define $c^T = \begin{bmatrix} c_1 & c_2 \end{bmatrix}$. With this substitution, (1.13) becomes

$$s = c^T z \tag{1.14}$$

where $z = \begin{bmatrix} z_1 & z_2 \end{bmatrix}^T$. By some control law, which will be discussed later, sliding mode can be ensured in finite time. During sliding mode $s(x,t) = 0$ implies

$$z_2 = -c_2^{-1}c_1z_1. \tag{1.15}$$

The above equation shows that during sliding mode, z_2 states are algebraically related to z_1 states and thus during sliding mode system order reduces by m, i.e by the number of inputs. Using (1.10) and (1.15)

$$\dot{z}_1 = (A_{11} - A_{12}c_2^{-1}c_1)z_1. \tag{1.16}$$

During sliding, the above equation decides dynamics; and surface parameters c_1 and c_2 should be selected based on system's performance specifications. For the design simplicity, c_2 is chosen as identity. By fixing the parameter c_2, the parameter c_1 can be designed by many different techniques like pole placement, eigen structure assignment, linear quadratic regulator (LQR) approach and Lyapunov approach.

1.1.2 Existence Condition for Sliding Mode

The objective of sliding mode control is to ensure sliding motion in finite time from an arbitrary initial condition. As we studied in the first order example where sliding surface is $s(x,t) = x(t)$, the sign of $s(x,t)$ and $\dot{s}(x,t)$ should be opposite to ensure finite time reaching. To ensure finite time reaching for general $n^{th}-$ order single input system the following conditions should be satisfied:

$$\lim_{s \to 0_+} \dot{s} < 0, \tag{1.17a}$$

$$\lim_{s \to 0_-} \dot{s} > 0. \tag{1.17b}$$

The above conditions precisely ensure that s and $\dot{s}$ should have opposite signs and due to this

$$s\dot{s} < 0. \tag{1.18}$$

The above condition is also known as the *reachability condition* in literature. The condition (1.18) ensures only asymptotic reaching on sliding surface. A stronger condition for finite time reaching is given as follows

$$s\dot{s} < -\eta|s|, \tag{1.19}$$

for some $\eta > 0$, known as η readability condition in literature. For a general $m-$input system, it is not necessary to ensure sliding mode on each discontinuity surface, sliding mode should exit on intersection of all discontinuity surfaces. These conditions ascertain that sliding surfaces remain attractive. Usually a control law which satisfies condition (1.19) is designed as

$$u(t) = -(c^T B)^{-1}(c^T A_r z(t) + Qsign(s)), \tag{1.20}$$

where

$$A_r = \begin{bmatrix} A_{11} & A_{12} \\ A_{21} & A_{22} \end{bmatrix} \tag{1.21}$$

and Q is an $m \times m$ diagonal matrix with positive entries which are chosen from the corresponding upper bounds of uncertainty. The above control law is derived such that the condition (1.19) is satisfied. Alternatively, a control law can be obtained by the so-called *reaching law* approach [51] in which the switching function dynamics are specified *a priori*. The following reaching laws were proposed in [51]:

- The constant rate reaching law:

$$\dot{s} = -Qsign(s) \tag{1.22}$$

- The constant plus proportional rate reaching law:

$$\dot{s} = -Ks - Qsign(s) \tag{1.23}$$

- The power rate reaching law:

$$\dot{s}_i = -\bar{Q}|s_i|^{\alpha_i} sign(s_i) \tag{1.24}$$
$$i = 12 \cdots m$$

where Q and K are diagonal matrices with positive elements, $\bar{Q}$ is a positive scalar, and $\alpha_i \in (0, 1)$. The reaching laws mentioned above can be used to obtain a control law and with appropriate choice of parameters it satisfies condition (1.19).

It is worth noting that the robustness properties of SMC is explored in several applications; [81, 18, 80, 66], to cite a few.

1.2 Discrete-Time Sliding Mode Control

Because of the flexibility of implementation, a large class of continuous systems are controlled by digital signal processors (DSPs) and high end microcontrollers. To analyze the effect of sampling time, discrete-time sliding mode control(DSMC) is well studied in the literature [89, 46, 87, 8, 44, 47, 12, 75]. In case of DSMC design, the control input is changed only at each sampling instant and the control effort is constant over the entire sampling period. Moreover, when the sliding mode control law is based on a switching function, similar to its continuous-time counterpart, the subsequent control would generally be unable to keep the states confined to the sliding surface. As a result, switching based DSMC can undergo only quasi-sliding mode (QSM), i.e., the system states would approach the sliding surface but would generally be unable to stay on it [46]. Thus, in general, QSM does not possess the invariance property found in continuous-time sliding mode. In [99, 12], it is shown that to achieve the sliding motion in discrete-time system, a discontinuous function in the control law is not necessary; equivalent control is sufficient to ensure sliding motion. However, to implement equivalent control law for an uncertain system, the exact value of uncertainty is needed. In practice, it is difficult to obtain the exact value of the uncertainty and this gives rise to a boundary layer around the sliding surface $s(k) = 0$. The width of the boundary layer depends on how the uncertainty is approximated. The width of boundary layer can also be reduced by incorporating a disturbance observer [89]; this approach ensures sufficiently small boundary layer for slowly varying disturbances or when the sampling frequency is large enough. In this Section, we present in brief a design methodology of DSMC. To motivate the idea of DSMC, we begin with a first order discrete-time system. Consider Euler's discretization of the system described in (1.1):

$$\frac{x(k+1)-x(k)}{\tau} = ax(k)+bu(k)+\rho(k).$$

where τ is the sampling time, $x(k) \in \mathbb{R}$, $u(k) \in \mathbb{R}$ and $\rho(k)$ accounts for the system uncertainty. The above equation can be rearranged in standard form as

$$x(k+1) = (1+a\tau)x(k)+\tau bu(k)+\tau\rho(k). \tag{1.25}$$

The objective is to design a control law to stabilize the above system. Consider a sliding surface defined as

$$s(k) \equiv c^T x(k), \tag{1.26}$$

where c^T is the parameter of surface. For the system in (1.25), let $c^T = 1$. In DSMC, the objective is to achieve $s(k) = 0$ in finite time. To achieve this objective Utkin [99, 98] proposed an equivalent control law which is derived by setting $s(k+1) = 0$.

This control law ensures one step reaching onto the sliding surface from any initial condition. For the system in (1.25) equivalent control becomes

$$u(k) = -(\tau b)^{-1}(1 + a\tau x(k) + \tau\rho(k)). \tag{1.27}$$

The above control law is derived by setting $s(k+1) = 0$. This control law contains uncertain terms and, therefore, it can not be implemented. As mentioned earlier, there are different approaches proposed by different researchers to estimate uncertainty in the system. In [12], the actual unknown disturbance is replaced by its average value. In [89], a delayed disturbance estimator is proposed by which the current unknown disturbance is replaced by its previous instant value. This approach gives better results for slowly varying disturbance. Without any disturbance compensation, the control law becomes

$$u(k) = -(\tau b)^{-1}(1 + a\tau x(k)). \tag{1.28}$$

The above control law results bound of $s(k)$ as

$$|s(k)| = |c^T x(k)| \leq \tau\rho_m, \tag{1.29}$$

where $|\rho(k)| \leq \rho_m$. It can be seen that with a small sampling period, the width of boundary layer around $s(k)$ becomes small. However, a small sampling period also increases control amplitude. It can be concluded that for an uncertain discrete-time system, the uncertainty gives rise to a boundary layer around $s(k) = 0$ and, therefore, only quasi-sliding mode is possible. A possible definition of discrete-time quasi-sliding mode is given as follows:

Definition 1.2.1 *[12](**Quasi Sliding Mode**). The Quasi-Sliding Mode is the motion in a predefined ε vicinity of the sliding surface $s(k) = 0$ such that the system trajectory, after entering this band, never abandons it, i.e $\|s(k)\| \leq \varepsilon$ where, the positive constant ε is called quasi-sliding mode band width.*

In what follows, a brief insight into designing a sliding surface for DSMC is presented. Consider an uncertain discrete-time system

$$x(k+1) = \Phi x(k) + \Gamma u(k) + \Gamma d(k), \tag{1.30}$$

where $x(k) \in \mathbb{R}^n$, $u(k) \in \mathbb{R}^m$. Here, n is the order of system and m is the number of inputs. For the above system, we assume that the pair (Φ, Γ) is controllable and the matrix Γ is of full rank. Then there exist an orthogonal transformation matrix T_r such that

$$T_r \Gamma = \begin{bmatrix} O \\ \Gamma_2 \end{bmatrix}, \tag{1.31}$$

where Γ_2 is an $m \times m$ nonsingular matrix. With this transformation the system in (1.30) can be represented as

$$z_1(k+1) = \Phi_{11}z_1(k) + \Phi_{12}z_2(k), \tag{1.32a}$$

$$z_2(k+1) = \Phi_{21}z_1(k) + \Phi_{22}z_2(k) + \Gamma_2 u(k) + \Gamma_2 d(k), \tag{1.32b}$$

where $\left[\, z_1(k)\; z_2(k) \,\right]^T = T_r x(k)$ and $z_1(k) \in \mathbb{R}^{n-m}$, $z_2(k) \in \mathbb{R}^m$. The sliding surface can be designed as

$$s(k) = c_1 z_1(k) + c_2 z_2(k), \tag{1.33}$$

where $s(k) \in \mathbb{R}^m$, $c_1 \in \mathbb{R}^{m\times(n-m)}$ and $c_2 \in \mathbb{R}^{m\times m}$. Suppose there exists a control law which enforces ideal sliding motion $s(k) = 0$, this implies

$$z_2(k) = -c_2^{-1} c_1 z_1(k).$$

By substituting $z_2(k)$ from the above equation into (1.32) it follows

$$z_1(k+1) = (\Phi_{11} - \Phi_{12} c_2^{-1} c_1) z_1(k).$$

To ensure stability of the sliding surface the above subsystem has to be stable. There are several techniques like pole placement, LQR, Lyapunov approach and so on which can be used to achieve problem specific objectives. Surface parameters c_1 and c_2 can be chosen to be either time-varying or nonlinear to meet the require objectives.

1.2.1 Existence of Discrete-Time Sliding Mode and Control Law

As argued earlier, a control law is supposed to bring sliding motion. The control law should be so designed that from any initial condition the state trajectory is forced on sliding surface $s(k) = 0$ in finite time. Direct discretization of continuous-time reachability gives only necessary conditions for discrete-time sliding mode. To obtain a necessary and sufficient condition for the existence of discrete-time sliding mode the following condition should satisfy:

$$|s(k+1)| < |s(k)|. \tag{1.34}$$

It has been established by Utkin in [98] that the discrete-time sliding mode can exist even without discontinuous control law, unlike continuous-time SMC. Furthermore, in the same work he proposed an equivalent control based control law which ensures ideal sliding mode for a completely deterministic system which also satisfies (1.34). There is another spool of thought to design control law which is based on direct discretization of continuous-time reaching laws [46, 47]. However, this approach does not ensure ideal sliding mode ($s(k) = 0$) even for completely deterministic systems as equivalent control does; and in [12] it is proved that this approach increases the width of the quasi-sliding mode band. This approach also induces *chattering* due to presence of discontinuous component which impedes the applicability of DSMC in many systems. For system (1.32), equivalent control can be obtained by setting $s(k+1) = 0$ as

$$u_{eq}(k) = -(c^T \Gamma_r)^{-1}(c^T \Phi_r z(k)) - d(k). \tag{1.35}$$

To implement the above control law, the exact value of disturbance $d(k)$ is needed. However, generally disturbance for any practical system is not known. To circumvent this difficulty, as mentioned earlier, there are different approaches by which an unknown disturbance is replaced (or estimated) by its average value [12], or the value at the previous sampling instant is used [89]. However, all the existing methods in literature can not ensure estimation of exact disturbance. Inability to know exact value of uncertainty/disturbance gives rise a boundary layer around $s(k) = 0$ and, in practice, the so-called quasi-sliding mode (QSM) results. Therefore, for an uncertain discrete-time system, the existence condition (1.34) should satisfy only outside the quasi-sliding mode band.

1.3 Literature Review and Motivation

This monograph aims to provide SMC based algorithms to ensure high performance in an uncertain environment. In SMC, the sliding surface decides dynamics, therefore it should be designed such that it addresses all requirements. One of the key requirements in many applications like robotics, electric drives, process control, vehicle and motion control is to have high performance in an uncertain environment. To enhance the performance of the system with sliding mode control algorithms, a time-varying switching surface is proposed by many researchers in [26, 11, 27, 14]. Majority of these solutions are for second order and third order systems. In [27], an algorithm based on moving switching line is presented for second order system to enhance the performance and to eliminate the reaching phase. However, in [11] it is shown that this method does not ensure complete insensitivity to external disturbances and model mismatches. This happens because switching line slope changes in discrete steps and between two consecutive steps the state trajectory is in the reaching phase, therefore the system loses invariance during the transient phase when the switching line slope is being changed. In [79], a strategy based on fuzzy logic is devised to change the parameters of switching surfaces of higher order systems. Some researchers proposed nonlinear surfaces to improve the performance. In [111, 50], proximate time optimal control is used to design the switching line for hard disc drive seek control applications. This approach is applicable to second-order systems only. In a paper [68], the authors proposed a nonlinear sliding surface and they also noted that for higher order system their approach becomes computationally intensive. Most of the existing solutions are applicable to second and third order system, although some are very important from the practical point of view (e.g., see [14]). To the best of authors' knowledge, a few results are devoted to enhance the performance of higher order systems. These observations motivate us to search for a better technique to ensure high performance in an uncertain environment.

To ensure high performance, system should settle quickly without any overshoot. It is well understood that a low overshoot can be achieved at the cost of high settling time. Low settling time is also necessary for a quick response. Thus, most of the design schemes make a tradeoff between these two transient performance indices, and the damping ratio is chosen as a fixed number. Notable exceptions exist, of course, in [70], the authors proposed a seminal idea of composite nonlinear feedback (CNF) for a class of second order systems subject to actuator saturation. The CNF uses a variable damping ratio to achieve high performance. The CNF control consists of a linear feedback law and a nonlinear law without any switching element. The linear part is designed for a small damping ratio to achieve a quick response. The nonlinear feedback is used to increase damping ratio as the output approaches the commanded target reference and thus overshoot is avoided. The CNF applies more control efforts when the output is nearer to the commanded target reference, resulting in better utilization of actuator capacity.

Subsequently, the CNF controller was extended for general higher order SISO and MIMO systems in [94, 93, 21, 49] for the state feedback and output feedback case. However, all these methods ensure performance only for perfectly known systems or when disturbance is constant. In [20], an enhanced CNF controller is proposed by adding integral action in the forward path. However, in [22], it is shown that integral action in forward path does not give robust performance for all types of disturbances. To solve this problem, in [22], a robust CNF controller is proposed based on constant disturbance estimation which is observed by an observer, and the effect of constant bias is compensated. In general, model uncertainty and disturbances are inevitable in actual applications, which would restrict the applicability of these results in practice. To be effective in practice along with the change of damping ratio, the controller should reject all kinds of disturbances.

Hence, to ensure robustness with high performance a nonlinear surface is proposed by the authors in [1, 3, 2, 7, 4] for different uncertain systems. The nonlinear surface changes system's closed loop damping ratio as the output approaches the setpoint. Initially, the nonlinear surface keeps the damping ratio to a low value to ensure quick response and as the output approaches the setpoint, the system is made highly damped to avoid overshoot. The nonlinear surface continuously changes the damping ratio of system from its initial low value to the final high value. A systematic linear matrix inequalities (LMIs) based algorithm is proposed to ease tuning of the parameters of the nonlinear surface. The next four chapters to follow discuss different nonlinear sliding surfaces to improve the performance for different types of system. For considering various performance requirements in sliding mode, the last two chapters, which are based on the authors' works in [63, 59, 64, 62], handle Lyapunov matrix based parametric approaches that utilize the well known linear control theories. For instance, the parameter uncertainties possibly *unmatched* can be systematically treated by the guaranteed cost control concept in sliding mode. The proposed methods eventually form a framework for multi-objective sliding mode design that comprises of several performance constraints such as pole-clustering, quadratic performance optimization, etc, which enhances the performance in sliding mode.

The main part of this monograph is organized as follows.

- The philosophy of the nonlinear sliding surface and how it changes the damping ratio is presented in the second chapter of the monograph . The systems dealt with in this chapter are uncertain linear and a class of uncertain nonlinear systems. To ease the synthesis of a nonlinear surface, the design of nonlinear surface is posed as a convex optimization problem in which a set of LMIs needs to be solved. It has been shown that the proposed nonlinear surface outperforms the different linear surfaces designed from different principles. Some important practical examples of linear and nonlinear systems are presented to verify the effectiveness of the scheme.

- The third chapter of the monograph discusses a high performance tracking controller for discrete systems [3]. A nonlinear sliding surface for general MIMO uncertain system is presented in this chapter. The proposed control law is based on the reaching law approach and using the concept of disturbance observer [89]. To overcome the requirement of measuring the complete state vector (as typically required by SMC), the control law uses multi-rate output feedback (MROF) [8] to relax this requirement. The disturbance/uncertainty is also measured by output samples only.
- Time delays constitute yet another aspect of systems, that needs special treatment. It is a well-recognized fact that the existence of time delay may produce poor performance or even instability [65]. The sliding mode control law formulation to improve the performance for an uncertain Linear Time Invariant (LTI) systems with time-delay forms the fourth chapter of the monograph. Generally, many systems have delays in their actuator circuitry. This would translate into input delay in the model. A nonlinear sliding surface is proposed in predicted state to achieve better transient response for a general discrete linear systems with input delay. A general uncertain input delay system is considered which contains both matched and unmatched uncertainties. The tracking problem is analyzed and ultimate boundedness of tracking error is proved.
- Chapter 5 of the monograph presents a new algorithm based on integral sliding mode to achieve high performance in an uncertain environment with limited control efforts (saturated actuator) [5, 4]. Actuator saturation issue is also considered and the stability of the system is ensured against saturated actuator. The integral sliding mode ensures invariance to matched disturbances right from the beginning without having any reaching phase [100, 19]. The control law comprises of two components: a nominal control part and a discontinuous control part. The nominal control part is designed based on composite nonlinear feedback (CNF) [70, 94, 21], which changes the damping ratio from its initial low value to final high value to achieve high performance. The algorithm presented solves the robustness issues of the CNF. The same algorithm is also evaluated through experiments on a laboratory based servo position control setup and compared with the CNF to prove the superiority of the presented algorithm.
- For incorporating multiple objectives in sliding mode, Chapter 6 of the monograph presents a multi-objective sliding mode design framework [63, 64, 62]. The proposed approach relies on formulating the Lyapunov matrix based constraints in LMIs imposed by several design objectives. The design objective represented by full order LMIs is shown to be also valid for the associated reduced order system. This allows the proposed approach to cope with various design objectives

based on the known results from the linear control theories such as robust control to unmatched parameter uncertainties, guaranteed control, pole-clustering problem, quadratic performance optimization, etc.

- Chapter 7 of the monograph discusses a complete sequence of SMC design in the discrete-time domain, by extending the results of Chapter 6 [59, 58]. It turns out that the discrete-time Lyapunov inequality of full order does generate all the stable sliding surfaces without loss of generality. Based on this, the quadratic performance optimization problem in sliding mode can be defined and systematically solved by LMIs. Also, to eliminate the assumption of full state availability, the multirate output feedback is combined and the upper bound of sliding function for quasi-sliding mode is explicitly estimated.

Chapter 2
High Performance Robust Controller Design Using Nonlinear Surface

2.1 Introduction

Many practical systems call for an improvement in transient performance along with the steady state accuracy. For example, many electro-mechanical, robotics and power converter systems require a quick response without any overshoot. It is a well understood fact that a low overshoot can be achieved at the cost of high settling time. However, a low settling time is also necessary for a quick response. Thus, most of the design schemes make a tradeoff between these two transient performance indices and the damping ratio is chosen as a fixed number. As explained in the first chapter, a variable damping ratio improves the system performance significantly. This chapter presents a method to design a nonlinear sliding surface for a linear uncertain system; and the method is also extended for a class of nonlinear system. A nonlinear sliding surface is designed by using the principle of composite nonlinear control [70, 94, 21]. Using a nonlinear sliding surface, the damping ratio of a system can be changed from its initial low value to final high value. The initial low value of damping ratio results in a quick response and the later high damping avoids overshoot. Thus the proposed surface ascertains the reduction in settling time without any overshoot. Furthermore, systems's damping ratio changes continuously as per the chosen function. Both regulator and tracking cases are considered in this chapter. The proposed approach inherits the robustness of SMC and delivers high performance due to change of damping ratio through the nonlinear sliding surface. During sliding mode, because of the order reduction, system response is unaffected by m poles. For a systems of order higher than 2, the damping ratio is specified by considering the contribution of dominant poles. However, non-dominant poles always affect the system response to some extent depending on their relative locations with respect to the dominant poles. Due to the order reduction property of SMC, m non-dominant poles will not contribute in the system response and thus, the performance specifications can be achieved more closely. The proposed nonlinear sliding surface achieves high performance and robustness unlike a sliding surface designed by assigning eigenvalues or by minimizing a quadratic index, which

B. Bandyopadhyay et al.: Sliding Mode Control Using Novel Sliding Surfaces, LNCIS 392, pp. 17–39.
springerlink.com

normally lead to a linear sliding surface. This chapter contains some results from [2] and several additional results.

The brief outline of this chapter is as follows. Section 2 describes the structure of the nonlinear sliding surface and its stability for the regulator case. It also gives an insight of change of damping ratio with the help of a third order system. Section 3 discusses the existence of the sliding mode by a new sliding mode control law. The tracking case is analyzed in Section 4. A possible generalization to a class of nonlinear systems is presented in Section 5. A modification in the sliding surface to eliminate the reaching phase is discussed in Section 6. Examples and simulation results are presented in Section 7 followed by the concluding Section 8.

2.2 Nonlinear Sliding Surface

This section discusses the design of a nonlinear sliding surface for the general SISO case with matched perturbations. Consider the system described by the following equation

$$\dot{x}(t) = Ax(t) + Bu(t) + \rho_d(x,t), \tag{2.1}$$

$$y(t) = C_1 x(t), \tag{2.2}$$

where $x(t) \in \mathbb{R}^n$, $u(t) \in \mathbb{R}$ and $y(t) \in \mathbb{R}$ are the state vector, input and output of the system, respectively. A, B and C_1 are the matrices of appropriate dimensions. $\rho_d(x,t)$ accounts for parametric perturbation or external disturbances, which is assumed to be *matched*, i.e., lie in the the range space of input matrix. Furthermore, only upper bound of $\rho_d(x,t)$ is known. The pair (A,B) is controllable and the matrix B is of full-column rank. Without loss of generality, the plant described by (2.1) and (2.2) can be transformed by some transformation matrix T_r into a regular form. Define $z := T_r x := \left[z_1(t) \; z_2(t) \right]^T$

The system in regular form can be described as

$$\dot{z}_1(t) = A_{11} z_1(t) + A_{12} z_2(t), \tag{2.3a}$$

$$\dot{z}_2(t) = A_{21} z_1(t) + A_{22} z_2(t) + B_2 u(t) + \rho_{d2}(z,t), \tag{2.3b}$$

$$y(t) = Cz(t), \tag{2.3c}$$

where $z_1(t) \in \mathbb{R}^{n-1}$, $z_2(t) \in \mathbb{R}$, $C = C_1 (T_r)^{-1}$, $\rho_{d2}(z,t) = T_r \rho_d(x,t)$ and B_2 is a scalar. Also, for the later usage, let us define a matrix

$$A_{reg} := \begin{bmatrix} A_{11} & A_{12} \\ A_{21} & A_{22} \end{bmatrix}. \tag{2.4}$$

Let the sliding surface for the system in regular form be defined as follows:

$$s(z,t) := c^T z(t) := \left[c_1 \; c_2 \right] \begin{bmatrix} z_1(t) \\ z_2(t) \end{bmatrix} := \left[F - \Psi(y) A_{12}^T P \;\; 1 \right] \begin{bmatrix} z_1(t) \\ z_2(t) \end{bmatrix}. \tag{2.5}$$

It is noted that F is chosen such that $(A_{11} - A_{12}F)$ has stable eigenvalues and the dominant poles have a low damping ratio. $\Psi(y)$ is a non-positive differentiable function in y, used to change the damping ratio. P is a positive definite matrix, which is chosen based on final damping ratio ζ_2 and also satisfies the Lyapunov equation

$$P(A_{11} - A_{12}F)^T + (A_{11} - A_{12}F)P = -W, \tag{2.6}$$

for some positive definite matrix W. Using this nonlinear sliding surface the damping ratio of the system is increased from its initial value ζ_1 to the final value ζ_2. The matrix P is chosen based on the final damping ratio ζ_2. The formulation to calculate the matrix P based on the required ζ_2 is discussed later in this section.

2.2.1 Choice of Nonlinear Function $\Psi(y)$

The nonlinear function $\Psi(y)$ is used to change the system's closed loop damping ratio from its initial low value to a final high value as the output varies from its initial value and approaches the setpoint(the origin for the regulator case). The nonlinear function should have the following two properties:

- It should change from 0 (or very small value) to -β as the output approaches the setpoint from its initial value, where $\beta > 0$.
- It should be differentiable with respect to y to ensure the existence of sliding mode.

One possible choice of $\Psi(y)$ which respects the above properties is as follows:

$$\Psi(y) = -\frac{\beta}{1 - e^{-1}}(e^{-(1-((y-y_0)/(-y_0))^2)} - e^{-1}), \tag{2.7}$$

where $y_0 = y(0)$, and β is used as a tuning parameter. This parameter contributes to decide the final damping ratio along with the matrix P. It should be noted that the choice of $\Psi(y)$ is not unique and any function with the above mentioned properties can be used. Another possible choice is as follows:

$$\Psi(y) = -\beta e^{-\bar{k}y^2}, \tag{2.8}$$

where $\bar{k}$ is a positive constant which should be of large value to ensure a small initial value of Ψ.

Stability of Sliding Surface

During the sliding mode (i.e., $s(z,t) = 0$), from (2.5),

$$z_2(t) = -Fz_1(t) + \Psi(y)A_{12}^T P z_1(t). \tag{2.9}$$

From (2.3) and (2.9), we can write the system during sliding mode as follows:

$$\dot{z}_1 = (A_{11} - A_{12}F + \Psi(y)A_{12}A_{12}^T P)z_1(t). \tag{2.10}$$

To prove the stability in sliding mode, it should be shown that the subsystem (2.10) is stable. The stability of (2.10) is proved in the following theorem.

Theorem 2.2.1. *If* $(A_{11} - A_{12}F)$ *is stable and* $\Psi(y)$ *is defined by (2.7) or (2.8), then subsystem in (2.10) is stable.*

Proof. Let a Lyapunov function for the system in (2.10) be defined as:

$$V(z) = z_1^T(t)Pz_1(t).$$

Then, it follows that

$$\begin{aligned}
\dot{V}(z) &= \dot{z_1}^T(t)Pz_1(t) + z_1^T(t)P\dot{z}_1(t)\\
&= z_1^T(t)(A_{11} - A_{12}F)^T Pz_1(t) + z_1^T(t)P(A_{11} - A_{12}F)z_1(t)\\
&\quad +2\Psi(y)z_1^T(t)PA_{12}A_{12}^T Pz_1(t)\\
&= z_1^T(t)\{(A_{11} - A_{12}F)^T P + P(A_{11} - A_{12}F)\}z_1(t) + 2\Psi(y)z_1^T(t)PA_{12}A_{12}^T Pz_1(t)\\
&= z_1^T(t)\{(A_{11} - A_{12}F)^T P + P(A_{11} - A_{12}F) + 2\Psi(y)PA_{12}A_{12}^T P\}z_1(t)\\
&= z_1^T(t)\{-W + 2\Psi(y)PA_{12}A_{12}^T P\}z_1(t).
\end{aligned}$$

Therefore, we have

$$\dot{V}(z) = z_1^T(t)\{-W + 2\Psi(y)MM^T\}z_1(t)$$

where $M := PA_{12} \in \mathbb{R}^{(n-1)}$. As $PA_{12}A_{12}^T P = MM^T \geq 0$ and the function $\Psi(y)$ is negative by definition, therefore matrix $2\Psi(y)MM^T$ is a negative semi definite. The matrix $-W$ is negative definite and the addition of a negative definite and a negative semidefinite matrix always results in a negative definite matrix. Therefore, we can write

$$\dot{V}(z) < 0,$$

which completes the proof. □

2.2.2 An Insight of Change in Damping Ratio

In the previous section, the structure of nonlinear sliding surface and the stability is discussed. The objective of this section is to build a comprehension that how the poles and damping ratio change with the corresponding change in the function Ψ during the sliding mode. To keep the simplicity of presentation, a third order single input system is used to describe how the damping ratio changes. However, the proposed nonlinear sliding surface can also be used with higher order plants. Our analysis closely follows to the work of Lin *et al.* [70] in which the concept of variable damping ratio was introduced for a second order deterministic system.

Let us assume, without loss of generality, that the system in (2.3) is in controllable canonical form. Suitable F can be chosen so that the matrix $A_{11} - A_{12}F$ becomes

Hurwitz and has a small damping ratio. Consider a system in controllable canonical form and, so thus,

$$A_{12} = \begin{bmatrix} 0 \\ 1 \end{bmatrix}. \tag{2.11}$$

Also, define

$$A_{11} - A_{12}F := \begin{bmatrix} 0 & 1 \\ -a_1 & -a_2 \end{bmatrix}, \tag{2.12}$$

where a_1 and a_2 are positive constants. Let us define the positive definite matrices P and W which satisfy (2.6) as

$$P := \begin{bmatrix} p_1 & p_2 \\ p_2 & p_3 \end{bmatrix} \tag{2.13}$$

and

$$W := \begin{bmatrix} w_1 & w_2 \\ w_2 & w_3 \end{bmatrix}. \tag{2.14}$$

By using the above values of matrices, (2.10) can be rewritten as

$$\dot{z}_1 = \begin{bmatrix} 0 & 1 \\ -a_1 & -a_2 \end{bmatrix} z_1(t) + \begin{bmatrix} 0 & 0 \\ \Psi p_2 & \Psi p_3 \end{bmatrix} z_1(t),$$

which can be simplified as

$$\dot{z}_1 = \begin{bmatrix} 0 & 1 \\ -a_1 + \Psi p_2 & -a_2 + \Psi p_3 \end{bmatrix} z_1(t). \tag{2.15}$$

Eigenvalues of the above closed loop system can be given as

$$\lambda_1 = \frac{-a_2}{2} + \frac{\Psi p_3}{2} + \frac{1}{2}\sqrt{(a_2 - \Psi p_3)^2 - 4a_1 + 4\Psi p_2}, \tag{2.16a}$$

$$\lambda_2 = \frac{-a_2}{2} + \frac{\Psi p_3}{2} - \frac{1}{2}\sqrt{(a_2 - \Psi p_3)^2 - 4a_1 + 4\Psi p_2}. \tag{2.16b}$$

To obtain the above equations in terms of elements of matrix W, the following formula can be easily derived by using (2.6) as

$$p_2 = \frac{w_1}{2a_1}, \tag{2.17}$$

$$p_3 = \frac{w_3}{2a_2} + \frac{w_1}{2a_1 a_2}. \tag{2.18}$$

As discussed earlier, the function Ψ changes from initial zero value to some negative value. As the function Ψ changes, this results in the change in location of closed loop poles. Now, it can be easily seen that, as $\Psi \to -\infty$, the closed loop poles in (2.16) can be written as

$$\lim_{\Psi \to -\infty} \lambda_1 = -\frac{a_2 w_1}{w_1 + a_1 w_3},$$
$$\lim_{\Psi \to -\infty} \lambda_2 = -\infty.$$

It can be concluded that during sliding mode, as the function Ψ changes from its initial value to $-\infty$, the closed loop system's damping ratio increases to ∞. For higher order system, a root locus based approach as suggested in [21] can be used to analyse the system. A reduced order model, which is formed by the use of only dominant poles, can also be used to analyse the higher order systems. Similarly, when the function Ψ changes from 0 to $-\beta$ corresponding increase in the damping ratio can also be proved.

2.2.3 LMI Based Tuning Algorithm

During the sliding mode, the dynamics of the system is decided by the subsystem (2.10). The subsystem is stable for any non-positive value of $\Psi(y)$. The function $\Psi(y)$ changes from 0 to $-\beta$ as output changes from its initial value to zero. For any intermediate value of $\Psi(y)$ the closed loop system (2.10) is also stable. As it is proved in the above section that the introduction of this function changes the damping ratio of system from its initial value ζ_1 to the final value ζ_2, where $\zeta_2 > \zeta_1$. When $\Psi(y) = 0$ at $t = 0$, the damping ratio is contributed by F which is designed for a low damping ratio. When the output reaches closer to the origin, $\Psi(y)$ contributes significantly to increase the damping ratio of the system. When the output reaches the origin, the steady state value of $\Psi(y)$ becomes $\Psi(y) = -\beta$; therefore, the subsystem (2.10) can be written as

$$\dot{z}_1(t) = (A_{11} - A_{12}F - \beta A_{12}A_{12}^T P)z_1(t), \tag{2.19}$$

which decides the final damping ratio. Thus, the parameter β and the matrix P should be so designed that the dominant poles of (2.19) have the desired damping ratio. The equation (2.19) can be rewritten as

$$\dot{z}_1(t) = (A_{11} - A_{12}(F + \beta A_{12}^T P))z_1(t). \tag{2.20}$$

Let the required gain be K_2 for the desired final damping ratio ζ_2, which may be computed by the pole placement technique. Then, it should hold that

$$K_2 = F + \beta A_{12}^T P$$

in order to realize the desired damping in (2.20), which can be equivalently expressed by

$$A_{12}^T P - \frac{K_2 - F}{\beta} = 0. \tag{2.21}$$

However, it may not be always feasible for the matrix P to satisfy the constraints (2.6) and (2.21), simultaneously. Therefore, one may relax the constraint (2.21) as follows:

$$\| H \| \leq \varepsilon \tag{2.22}$$

for a sufficiently small $\varepsilon > 0$, where

$$H := A_{12}^T P - \frac{K_2 - F}{\beta}. \tag{2.23}$$

It is noted that the above nonlinear inequality can be converted into a linear inequality by using Schur complement as follows:

$$\begin{bmatrix} \varepsilon I & H \\ H^T & \varepsilon I \end{bmatrix} > 0. \tag{2.24}$$

Therefore we can cast the optimization problem as follows:

Choose P to minimize ε such that

$$P > 0, \tag{2.25}$$

$$P(A_{11} - A_{12}F)^T + (A_{11} - A_{12}F)P < 0 \tag{2.26}$$

$$\begin{bmatrix} \varepsilon I & H \\ H^T & \varepsilon I \end{bmatrix} > 0. \tag{2.27}$$

The above LMIs can be easily solved to minimize ε. It should be noted that the freedom to choose the parameter β can also be used to minimize the objective function. This approach automates the tuning procedure and reduces the number of manual iterations required for tuning.

Remark 2.2.1. *In (2.5), the matrix gain F is designed for a low damping ratio using pole placement technique. For higher order systems formulas of damping ratio and settling time are not defined explicitly in terms of system parameters. However, two complex dominant poles with the desired damping ratio and settling time can be chosen and the remaining poles can be placed away from the dominant poles to reduce its effect on the system response. The dominant pole concept is also widely used in the CNF literature [94, 49] to specify the damping ratio of higher order SISO and MIMO systems.*

2.3 Existence of Sliding Mode

The sliding surface discussed in the previous section is nonlinear and surface parameters are changing at every instant. A control law should be chosen in such a way that from any initial condition, the system trajectory is attracted towards the sliding surface and then slides along the surface. The existence of control law is discussed in the following theorem.

Theorem 2.3.1. *The control law*

$$u = -B_2^{-1}(c^T A_{reg}z + Ks + Qsign(s) - \frac{d\Psi(y)}{dt}A_{12}^T Pz_1(t)) \tag{2.28}$$

enforces the trajectory of (2.3) to move from any initial condition to the sliding surface in finite time and thereafter to remain on it. In the above control law the matrix Q is chosen from the maximum bounds of the uncertainty as follows:

$$Q \geq (\rho_{d2}(z,t))_{max}. \tag{2.29}$$

Proof. For a quadratic function, $V_1 = 0.5s^T s$, it may be shown that

$$\begin{aligned}\dot{V}_1 &= s^T[F\dot{z}_1 - \frac{d\Psi(y)}{dt}A_{12}^T Pz_1(t) - \Psi(y)A_{12}^T P\dot{z}_1 + \dot{z}_2],\\ &= s^T[(F - \Psi(y)A_{12}^T P)\dot{z}_1 - \frac{d\Psi(y)}{dt}A_{12}^T Pz_1(t) + A_{21}z_1(t) + A_{22}z_2(t) + B_2u\\ &\quad +\rho_{d2}(z,t)].\end{aligned}$$

By using (2.3) and (2.4), it leads to

$$\dot{V}_1 = s^T[c^T A_{reg}z + B_2u + \rho_{d2}(z,t) - \frac{d\Psi(y)}{dt}A_{12}^T Pz_1(t)].$$

Using the control law (2.28), it can be seen that

$$\begin{aligned}\dot{V}_1 &= s^T[c^T A_{reg}z + \rho_{d2}(z,t) - \frac{d\Psi(y)}{dt}A_{12}^T Pz_1(t) - c^T A_{reg}z - Ks\\ &\quad -Qsign(s) + \frac{d\Psi(y)}{dt}A_{12}^T Pz_1(t)]\\ &= s^T[-Ks - Qsign(s) + \rho_{d2}(z,t)],\end{aligned}$$

from (2.29), it implies $\dot{V}_1 < 0$. This completes the proof. □

It has been proved that from any initial condition, the state trajectory is forced towards the sliding surface by the control law (2.28) and the trajectory remains thereafter. It should be noted that the state trajectory hits the surface in finite time by the control law (2.28). After hitting the sliding surface, the trajectory will slide along the surface and the system is invariant against parameter variations and external disturbances satisfying the matching condition.

Remark 2.3.1. *The control law (2.28) requires the total derivative of the output For many systems derivative of output can be calculated in the form of state (i.e, when input and output subspaces are orthogonal to each other, due to this, for SISO case* $C_1 * B = 0$*). When it does not exist explicitly, it can be easily constructed by robust differentiator proposed by Levant [69] or Utkin [97]. It is worth noting that recently a large number of practical applications have witnessed the efficient use of these techniques to obtain the derivative of a signal.*

2.4 Robust Tracking Controller Based on Nonlinear Sliding Surface

This section discusses the tracking case for uncertain MIMO plant with matched perturbation. Consider the system described by the following equations

$$\dot{z}_1(t) = A_{11}z_1(t) + A_{12}z_2(t), \tag{2.30a}$$
$$\dot{z}_2(t) = A_{21}z_1(t) + A_{22}z_2(t) + B_2u(t) + \rho_{d2}(z,t), \tag{2.30b}$$
$$y(t) = Cz(t). \tag{2.30c}$$

where $z_1(t) \in \mathbb{R}^{n-m}$, $z_2(t) \in \mathbb{R}^m$ and $y(t) \in \mathbb{R}^p$, and $\rho_{d2}(z,t)$ represents disturbances or model mismatch with the known bounds. B_2 is an $m \times m$ matrix and all the other matrices are of appropriate dimensions. The above equation is similar to (2.3) but with different dimensions.

Let us define sliding surface for the system in (2.30) as

$$s(z,t) := \begin{bmatrix} F - \Psi(y,r)A_{12}^T P \; I_m \end{bmatrix} \begin{bmatrix} e_1(t) \\ e_2(t) \end{bmatrix}, \tag{2.31}$$

where

$$e_1(t) = z_1(t) - z_{1d}(t), \tag{2.32a}$$
$$e_2(t) = z_2(t) - z_{2d}(t), \tag{2.32b}$$

I_m is an $m \times m$ identity matrix and $z_d(t) = [z_{1d}^T(t) \; z_{2d}^T(t)]^T$ is the desired trajectory. F is chosen such that $(A_{11} - A_{12}F)$ has stable eigenvalues and the dominant poles are with a small damping ratio. $\Psi(y,r)$ is an $m \times m$ diagonal matrix with non-positive nonlinear entries depending on the output, and used to change the damping ratio. P is an $(n-m) \times (n-m)$ positive definite matrix satisfying the Lyapunov equation as in the regulator case discussed in Section 2.2.

2.4.1 *Selection of Matrix of Nonlinear Functions* $\Psi(y,r)$

$\Psi(y,r)$ is used to change the system's closed loop damping ratio from its initial low value to final high value as the output approaches from its initial value to the setpoint. It is chosen such that its diagonal elements $\Psi(y,r)_i$ change from 0 to -β_i ($\beta_i > 0$) in a nonlinear way. One possible choice of $\Psi(y,r)_i$ is as follows:

$$\Psi(y,r) = \begin{bmatrix} \Psi(y,r)_1 & \dots & 0 \\ \vdots & \ddots & \vdots \\ 0 & \dots & \Psi(y,r)_m \end{bmatrix}, \tag{2.33}$$

where

$$\Psi(y,r)_i = -\frac{\beta_i}{1-e^{-1}}(e^{-(1-(y-y_0)/(r-y_0))^2} - e^{-1}),\ i = 1,2,\cdots,m, \tag{2.34}$$

where $y_0 = y(0)$, and β_i is used as a tuning parameter which determines the final damping ratio together with P. Recall that function $\Psi(y,r)_i$ should have the two properties:

- It should change from 0 to -β_i as output approaches from its initial value to setpoint, where $\beta_i > 0$.
- It should be differentiable with respect to y.

It is noted that the choice of $\Psi(y,r)$ is not unique and any function with the above mentioned property can be used.

2.4.2 *Analysis in Sliding Mode*

During sliding mode $s(z,t) = 0$, it can be seen that, from (2.31),

$$e_2(t) = -Fe_1(t) + \Psi(y,r)A_{12}^T Pe_1(t). \tag{2.35}$$

Then, from (2.30) and (2.35), it leads to

$$\dot{e_1}(t) = (A_{11} - A_{12}F + A_{12}\Psi(y,r)A_{12}^T P)e_1(t) + g(t) \tag{2.36}$$

where $g(t) = A_{11}z_{1d}(t) + A_{12}z_{2d}(t) - \dot{z}_{1d}(t)$. Suppose that the desired trajectory $z_d(t)$ is consistently generated by using the system model [78]; then there exist some control u_d such that

$$\dot{z}_{1d}(t) = A_{11}z_{1d}(t) + A_{12}z_{2d}(t). \tag{2.37}$$

Using (2.36) and (2.37), it holds that

$$\dot{e}_1(t) = (A_{11} - A_{12}F + A_{12}\Psi(y,r)A_{12}^T P)e_1(t). \tag{2.38}$$

Now, for showing the stability of sliding mode, a theorem is introduced in the following.

Theorem 2.4.1. *If $(A_{11} - A_{12}F)$ is stable and $\Psi(y,r)$ is defined by (2.33), then the subsystem in (2.38) is stable.*

Proof. Let a Lyapunov function for system in (2.38) be defined as

$$V(e) = e_1(t)^T Pe_1(t).$$

Using (2.38) and rearranging, it can be shown that

$$
\begin{aligned}
\dot{V}(e) &= \dot{e}_1^T(t)Pe_1(t) + e_1^T(t)P\dot{e}_1(t) \\
&= e_1^T(A_{11} - A_{12}F)^T Pe_1 + e_1^T P(A_{11} - A_{12}F)e_1 \\
&\quad + 2e_1^T PA_{12}\Psi(y,r)A_{12}^T Pe_1, \\
&= e_1^T(t)\{(A_{11} - A_{12}F)^T P + P(A_{11} - A_{12}F)\}e_1(t) + \\
&\quad 2e_1^T(t)PA_{12}\Psi(y,r)A_{12}^T Pe_1(t) \\
&= e_1^T(t)\{(A_{11} - A_{12}F)^T P + P(A_{11} - A_{12}F) + \\
&\quad 2PA_{12}\Psi(y,r)A_{12}^T P\}e_1(t) \\
&= e_1^T(t)\{-W + 2PA_{12}\Psi(y,r)A_{12}^T P\}e_1(t).
\end{aligned}
$$

Let us define $M = e_1(t)^T PA_{12}$. Here M is a row vector. With this substitution, $\dot{V}(e)$ can be simplified as

$$\dot{V}(e) = -e_1^T We_1 + 2M\Psi(y,r)M^T.$$

Since $\Psi(y,r) < 0$ and $W > 0$, it leads to

$$\dot{V}(e) < 0.$$

This completes the proof. □

Existence of sliding mode is proved in the following theorem.

Theorem 2.4.2. *The control law*

$$u = -B_2^{-1}(-c^T\dot{z}_d + c^T A_{reg}z + Ks + Qsign(s) - \frac{d\Psi(y)}{dt}A_{12}^T Pz_1(t)) \tag{2.39}$$

enforces the trajectory of the system described by (2.30) to move from any initial condition to the sliding surface in finite time and thereafter to remain on it, where $c^T = \begin{bmatrix} F - \Psi(y,r)A_{12}^T P \; I_m \end{bmatrix}$,

$$A_{reg} = \begin{bmatrix} A_{11} & A_{12} \\ A_{21} & A_{22} \end{bmatrix}, \tag{2.40}$$

and the matrix Q is chosen from the maximum bound of the uncertainty as follows:

$$Q \geq (\rho_{d2}(z,t))_{max}. \tag{2.41}$$

Proof. Let a Lyapunov function for the system described by (2.30) be given as

$$V_1 = 0.5s^T s.$$

Using the definition of error from (2.32), it follows that

$$\begin{aligned}\dot{V}_1 &= s^T \left[F\dot{e}_1 - \frac{d\Psi(y)}{dt}A_{12}^T P e_1 - \Psi(y)A_{12}^T P\dot{e}_1 + \dot{e}_2\right] \\ &= s^T \left[(F - \Psi(y)A_{12}^T P)\dot{z}_1 - \frac{d\Psi(y)}{dt}A_{12}^T P z_1(t) + A_{21}z_1(t) + A_{22}z_2(t)\right. \\ &\left. + B_2 u + \rho_{d2}(z,t) - c^T \dot{z}_d\right],\end{aligned}$$

which gives, by using (2.30) and (2.40),

$$\dot{V}_1 = s^T \left[c^T A_{reg} z + B_2 u + \rho_{d2} - \frac{d\Psi(y)}{dt}A_{12}^T P z_1(t) - c^T \dot{z}_d\right].$$

By using control law (2.39), it can be seen that

$$\begin{aligned}\dot{V}_1 &= s^T \left[c^T A_{reg} z + \rho_{d2}(z,t) - \frac{d\Psi(y)}{dt}A_{12}^T P z_1(t) - c^T A_{reg} z - Ks\right. \\ &\left. - Qsign(s) + \frac{d\Psi(y)}{dt}A_{12}^T P z_1(t) + c^T \dot{z}_d - c^T \dot{z}_d\right] \\ \dot{V}_1 &= s^T \left[-Ks - Qsign(s) + \rho_{d2}(z,t)\right].\end{aligned}$$

Thus, with (2.41), it is easy to show that

$$\dot{V}_1 < 0.$$

This completes the proof. □

2.5 Nonlinear Surface for a Class of Nonlinear Uncertain Systems

A possible generalization to a class of nonlinear uncertain systems is presented in this section. Consider a system described by

$$\dot{x}_i = x_{i+1} \quad i = 1 \ldots n-1 \tag{2.42}$$

$$\dot{x}_n = f(x,u,t) + B_2 u, \tag{2.43}$$

where $x(t) = [x_1 \; x_2 \; \ldots \; x_n]^T$ represents the state vector, and $f(x,u,t)$ is an uncertain smooth function which satisfy the classical condition for the existence and uniqueness. Furthermore, the function satisfies

$$|f(x,u,t)| \leq Q_1 + Q_2|x(t)|, \tag{2.44}$$

where Q_1 and Q_2 are positive constants. The above definition relaxes the precise modeling of the plant. The theory of nonlinear sliding surface developed in the previous sections can be used to improve the performance of this nonlinear system. The surface proposed in (2.5) can be used for the above nonlinear system and corresponding submatrices can be defined as follows:

$$A_{11} = \begin{bmatrix} 0 & I_{n-2} \\ 0 & 0 \end{bmatrix}, \; A_{12} = \begin{bmatrix} 0 \\ 0 \\ \vdots \\ 1 \end{bmatrix}, \; A_{21} = [0, \cdots, 0], \; A_{22} = 0.$$

The other parameters of the nonlinear sliding surface can be found as per the discussion in Section 2.2. The control law (2.28) with

$$Q = Q_1 + Q_2 |x(t)| \tag{2.45}$$

ensures the existence of sliding mode for the system in (2.42) and (2.43). It is also straightforward to design a tracking controller for this nonlinear system as discussed in the previous section.

2.6 Elimination of the Reaching Phase

The initial phase when the system trajectory has not reached (or intercepted) the sliding surface is called the *reaching phase.* Although the control law ascertain finite time reaching on sliding surface from any initial condition, during the reaching phase even a matched perturbations can influence the system response. Therefore, the elimination of reaching phase ensures invariance property of SMC from the beginning. In this section we explain how the reaching phase can be eliminated with this nonlinear sliding surface for SISO regulator case and we also note that results presented in section can be easily extended for MIMO case. We adopt a technique presented in [109] , to eliminate the reaching phase, in the proposed nonlinear sliding surface framework. To eliminate the reaching phase, we modify the nonlinear sliding surface in (2.5) as

$$\bar{s}(z,t) = c^T \left(z(t) - z(0)e^{-\lambda t}\right) \tag{2.46}$$

$$= \begin{bmatrix} c_1 & c_2 \end{bmatrix} \begin{bmatrix} z_1(t) - z_1(0)e^{-\lambda t} \\ z_2(t) - z_2(0)e^{-\lambda t} \end{bmatrix} \tag{2.47}$$

$$= \begin{bmatrix} F - \Psi(y) A_{12}^T P & 1 \end{bmatrix} \begin{bmatrix} z_1(t) - z_1(0)e^{-\lambda t} \\ z_2(t) - z_2(0)e^{-\lambda t} \end{bmatrix}, \tag{2.48}$$

where λ is a positive constant and $z(0) = [z_1(0) \;\; z_2(0)]^T$ is the initial condition vector. The above augmented nonlinear surface satisfies $\bar{s}(z,t) = 0$ from the beginning. It should be noted that the effect of the term $z(0)e^{-\lambda t}$ vanishes with time and $\bar{s}(z,t)$ approaches $s(z,t)$. Therefore, the performance of the system in sliding mode is largely decided by $s(z,t)$ rather than $\bar{s}(z,t)$. Stability of this system can be proved in a similar way as discussed earlier because the term $z(0)e^{-\lambda t}$ goes to zero asymptotically. SMC law can be designed as

$$u = -B_2^{-1}(c^T A_{reg} z + Ks + Qsign(s) - \frac{d\Psi(y)}{dt} A_{12}^T P z_1(t) - \\ \frac{d\Psi(y)}{dt} A_{12}^T P z_1(0) e^{-\lambda t} + c^T z(0) \lambda e^{-\lambda t}). \tag{2.49}$$

With the above control law, the existence condition of sliding mode can be easily proved.

2.7 Example and Simulation Results

In this section, we illustrate two applications to show effectiveness of the proposed nonlinear sliding surface. First, position control DC motor, which is modeled as linear system with uncertainty, is considered. Secondly, position control of stepper motor, which is modeled by a set of nonlinear equations, is illustrated.

2.7.1 *Example 1: DC Motor Control*

We begin with position control of DC motor based on the armature controlled scheme. The control objectives are as follows:

- System should have low overshoot.
- System should have short settling time so it settles quickly.
- Controlled system should be robust against disturbances.

As per the specifications of DC motor given in [95] with load inertia J_L= 0.0012kg-m^2, the system parameters for (2.1) are given as follows:

$$A = \begin{bmatrix} 0 & 1 & 0 \\ 0 & -0.694 & 112.36 \\ 0 & -161.8 & -1500 \end{bmatrix}, \; B = \begin{bmatrix} 0 \\ 0 \\ 200 \end{bmatrix}, \; C_1 = \begin{bmatrix} 1 & 0 & 0 \end{bmatrix}.$$

And, the disturbance is assumed by

$$\rho_d(x,t) = \begin{bmatrix} 0 & 0 & 20sin(10t) \end{bmatrix}^T$$

Note that x_1 is the angular position, x_2 is the angular velocity, and x_3 is the armature current. Also, the control input u is the armature voltage.

Step 1: Transform the system into the regular form by appropriate T_r matrix. The above system is already in regular form therefore $T_r = I$ and other submatrices can be found as follows:

$$z_1(t) = \begin{bmatrix} x_1 & x_2 \end{bmatrix}^T, \; z_2(t) = x_3, \; \rho_{d2}(x,t) = 20sin(10t),$$

$$A_{11} = \begin{bmatrix} 0 & 1 \\ 0 & -0.694 \end{bmatrix}, \; A_{12} = \begin{bmatrix} 0 \\ 112.36 \end{bmatrix}, \; A_{21} = \begin{bmatrix} 0 & -161.8 \end{bmatrix}, \; A_{22} = -1500, \; B_2 = 200.$$

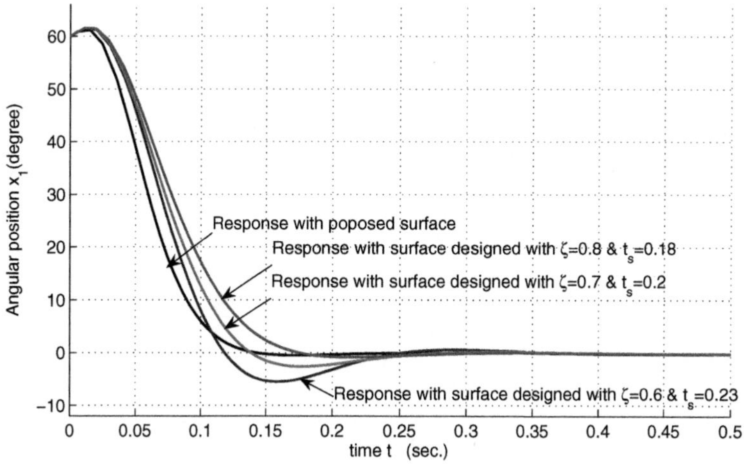

Fig. 2.1 Response of angular position x_1 with different sliding surfaces

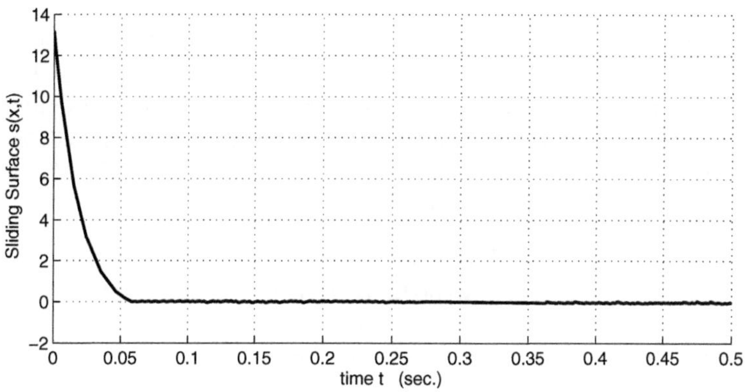

Fig. 2.2 Evolution of nonlinear sliding surface with time

Step 2: Design the nonlinear sliding surface. The nonlinear sliding surface is composed of a linear and a nonlinear term. Initially nonlinear term is zero, therefore the linear term decides initial damping ratio (ζ_1) and settling time.

Sliding surface can be given by the following equation

$$s = c_1 z_1(t) + c_2 z_2(t) \tag{2.50}$$

As per the discussion in Section 2.2, let $c_2 = 1$, and c_1 is made of two components linear and nonlinear.

$$c_1 = F - \Psi(y) A_{12}^T P.$$

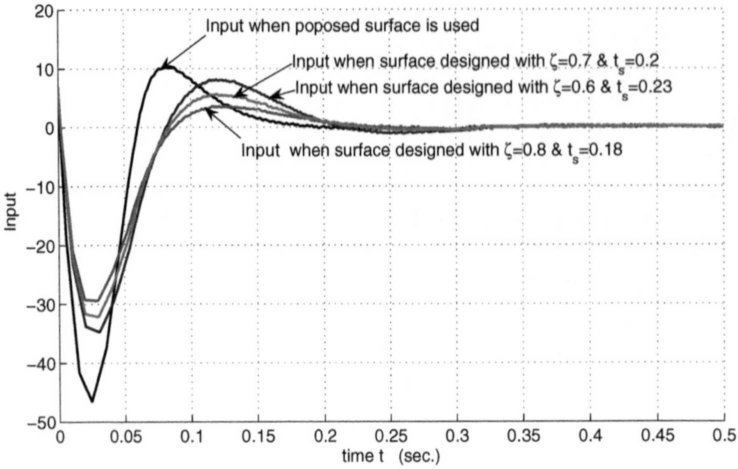

Fig. 2.3 Plot of input with different linear surfaces and nonlinear sliding surface

F is designed for low damping ratio. For initial settling time $t_{s1} = 0.23sec.$ and initial damping ratio $\zeta_1 = 0.5$, matrix F can be found by pole placement approach as $F = \begin{bmatrix} 10.76 \; 0.303 \end{bmatrix}$. Let the required final damping ratio be $\zeta_2 = 0.87$ and final settling time $t_{s2} = 0.13$. By following the procedure given in Section 2.2.3, the required gain matrix K_2 can be computed as $K_2 = \begin{bmatrix} 10.76 \; 0.53 \end{bmatrix}$. For solving LMIs (2.25), (2.26) and (2.27) for P with $\beta = 7.9$ gives one possible solution of P as follows:

$$P = 10^{-3} \begin{bmatrix} 0.0004 & -0.0074 \\ -0.0074 & 0.2631 \end{bmatrix}.$$

Also, function $\Psi(y)$ is designed with $\beta = 7.9$ as follows:

$$\Psi(y) = -\frac{7.9}{1-e^{-1}}(e^{-(1-((y-y_0)/(-y_0))^2)} - e^{-1}).$$

Therefore, the resulting sliding function is given by

$$s = \begin{bmatrix} F - \Psi(y)A_{12}^T P & 1 \end{bmatrix} \begin{bmatrix} z_1(t) & z_2(t) \end{bmatrix}^T. \tag{2.51}$$

Step 3: Design of control law. Control law

$$u = -B_2^{-1}(c^T Az + Ks + Qsign(s) - \frac{d\Psi(y)}{dt} A_{12}^T P z_1(t)) \tag{2.52}$$

can be defined by the values:

$$K = 36,\ Q = 29.5,\ x_0 = \begin{bmatrix} 60^\circ & 3 & 1 \end{bmatrix}^T,$$

where Q is chosen by considering maximum bound of uncertainty such that $(\rho_{d2})_{max} = 20$.

Design of linear sliding surface. Performance of the proposed nonlinear sliding surface is compared with the controller designed with different linear sliding surfaces. System output response is plotted for different linear surfaces and the proposed nonlinear sliding surface. Parameters of linear sliding surface(ζ and t_s) are chosen based on the values of these parameters taken at different instant when nonlinear sliding surface is used. When nonlinear sliding surface is used, poles of the closed loop system changes as output approaches to the origin. This changes the damping ratio and settling time from its initial values ζ_1 and t_{s1} to final values ζ_2 and t_{s2}. During the course of change, at different instant system has different damping ratio and settling time. Following three different linear sliding surfaces are designed based on the damping ratio and settling time occurred at different instant when nonlinear sliding surface is used.
(1) Case-1 Linear surface-1 with $\zeta = 0.6, t_s = 0.23$
(2) Case-2 Linear surface-2 with $\zeta = 0.7, t_s = 0.2$
(3) Case-3 Linear surface-3 with $\zeta = 0.8, t_s = 0.18$
Responses obtained by different linear surfaces are compared with that obtained by the nonlinear sliding surface. It should be noted that all these values of ζ and t_s occur at different instant when nonlinear sliding surface is used. Fixed sliding surface with given settling time and damping ratio is designed by pole placement approach.

$$s_i = c_i^T z = \left[F_i \; 1\right] \left[z_1(t) \; z_2(t)\right]^T . \tag{2.53}$$

In the above equation F_i is computed by pole placement approach for each of the three cases mentioned above with corresponding damping ratio and settling time. Control law for each of the three cases is given as follows:

$$u_i = -B_2^{-1}(c_i^T Az + Ks_i + Q sign(s_i)), \tag{2.54}$$

where the parameters Q and K are the same as (2.52). The system is simulated with the same initial condition and disturbance for different linear sliding surfaces.

In Fig. 2.1, response of angular position x_1 is plotted by using nonlinear sliding surface and different linear sliding surfaces. The plot clearly shows that with nonlinear sliding surface performance improves significantly. Furthermore it shows

Table 2.1 Comparison of performance obtained by proposed nonlinear sliding surface against linear sliding surfaces with different ζ

Type of sliding surface	Peak overshoot(%)	Settling Time t_s(sec.)
Linear sliding surface with ζ=0.6 and $t_s = 0.23$	9.33	0.23
Linear sliding surface with ζ=0.7 and $t_s = 0.2$	4.42	0.26
Linear sliding surface with ζ=0.8 and $t_s = 0.18$	1.42	0.27
Proposed nonlinear sliding surface	0.8	0.13

very clearly that peak overshoot and settling time both can be minimized simultaneously with nonlinear sliding surface. Detailed comparison for different performance parameters are given in Table 2.1. Fig. 2.2 shows evolution of nonlinear switching function with time. Fig. 2.3 shows the plot of input, it can be seen that due to nonlinear sliding surface gain increases as state reaches nearer to origin and thus settling time improves. Table 2.1 shows minimum improvement in settling time is 50% with negligible overshoot.

2.7.2 *Example 2: Stepper Motor Position Control*

In this subsection a high performance robust position tracking controller for permanent magnet (PM) stepper motor is presented by the proposed theory. To this end, let us review briefly the modeling of stepper motor.

A brief review of PM stepper motor model: Following the developments in [23] and [56], the mathematical model of stepper motor can be given based on the conservation of energy as follows:

$$\begin{aligned}
\frac{di_a}{dt} &= \frac{-r_a i_a + k_m \omega sin(n_r\theta) + v_a}{L}, \\
\frac{di_b}{dt} &= \frac{-r_b i_b - k_m \omega cos(n_r\theta) + v_b}{L}, \\
\frac{d\omega}{dt} &= \frac{-i_a k_m sin(n_r\theta) + i_b k_m cos(n_r\theta) - B_f\omega - \tau_l}{J}, \\
\frac{d\theta}{dt} &= \omega,
\end{aligned} \tag{2.55}$$

where winding-1 resistance r_a, rotor moment of inertia J, load torque τ_l, motor torque constant k_m, winding-2 resistance r_b, winding inductance L, number of rotor teeth n_r, viscous friction coefficient B_f, winding-1 current i_a, winding-2 current i_b, angular position θ, angular velocity ω, input voltage to winding-1 v_a and input voltage to winding-2 v_b. A nonlinear transformation, known as the direct-quadrature can be used to transform these equations into a form which is more suitable for designing nonlinear controllers. The direct-quadrature form is defined as follows:

$$\begin{bmatrix} i_d \\ i_q \\ w \\ \theta \end{bmatrix} = \begin{bmatrix} cos(n_r\theta) & sin(n_r\theta) & 0 & 0 \\ -sin(n_r\theta) & cos(n_r\theta) & 0 & 0 \\ 0 & 0 & 1 & 0 \\ 0 & 0 & 0 & 1 \end{bmatrix} \begin{bmatrix} i_a \\ i_b \\ w \\ \theta \end{bmatrix}. \tag{2.56}$$

In addition, we define the new inputs as

$$\begin{bmatrix} v_d \\ v_q \end{bmatrix} = \begin{bmatrix} cos(n_r\theta) & sin(n_r\theta) \\ -sin(n_r\theta) & cos(n_r\theta) \end{bmatrix} \begin{bmatrix} v_a \\ v_b \end{bmatrix}. \tag{2.57}$$

Suppose that J_L and B_L represent load inertia and viscous friction coefficient of the load , respectively. Then, load can be represented by

$$\tau_l = J_L \frac{d\omega}{dt} + B_L \omega. \tag{2.58}$$

Using (2.56), (2.57) and (2.58), the system equations can be simplified as

$$\dot{z_1}(t) = -k_1 z_1(t) + n_r z_2(t) z_3(t) + u_1(t) \tag{2.59a}$$
$$\dot{z_2}(t) = -k_1 z_2(t) - n_r z_1(t) z_3(t) - k_3 z_3(t) + u_2(t) \tag{2.59b}$$
$$\dot{z_3}(t) = k_4 z_2(t) - k_5 z_3(t) \tag{2.59c}$$
$$\dot{z_4}(t) = z_3(t) \tag{2.59d}$$

where

$$z_1(t) = i_d,\ z_2(t) = i_q,\ z_3 = \omega,\ z_4 = \theta,$$
$$k_1 = \frac{r}{L},\ k_3 = \frac{k_m}{L},\ k_4 = \frac{k_m}{J + J_L},\ k_5 = \frac{B_f + B_L}{J + J_L},$$
$$u_1 = v_d/L,\ u_2 = v_q/L.$$

Let the auxiliary inputs be defined as

$$\bar{u}_1(t) = u_1(t) + n_r z_3(t) z_2(t), \tag{2.60}$$
$$\bar{u}_2(t) = u_2(t) - n_r z_3(t) z_1(t). \tag{2.61}$$

Then, a linear model can be obtained by feedback linearization technique as follows:

$$\begin{bmatrix} \dot{z_1}(t) \\ \dot{z_2}(t) \\ \dot{z_3}(t) \\ \dot{z_4}(t) \end{bmatrix} = \begin{bmatrix} -k_1 & 0 & 0 & 0 \\ 0 & -k_1 & -k_3 & 0 \\ 0 & k_4 & -k_5 & 0 \\ 0 & 0 & 1 & 0 \end{bmatrix} \begin{bmatrix} z_1(t) \\ z_2(t) \\ z_3(t) \\ z_4(t) \end{bmatrix} + \begin{bmatrix} 1 & 0 \\ 0 & 1 \\ 0 & 0 \\ 0 & 0 \end{bmatrix} \begin{bmatrix} \bar{u}_1(t) \\ \bar{u}_2(t) \end{bmatrix},$$

which is represented in standard form

$$\dot{z}(t) = A z(t) + B \bar{u}(t). \tag{2.62}$$

For simulation, the physical parameters of stepper motor are summarized in Table 2.2. Also, the parameters for load are as follows:

J_L= 11.295e-6 Kg-m-sec^2, B_L=0.0013 Nm/rad/s.

Table 2.2 Stepper Motor Parameters

Parameter	Value
Winding Resistance	23 Ω
Winding Inductance	256e-3H
Rotor Inertia	1.1295e-4 Kg-m-sec^2
Friction Coefficient	0.0013 Nm/rad/s

Taking into account of all these numerical values, the system matrices of (2.62) are as follows:

$$A = \begin{bmatrix} -89.83 & 0 & 0 & 0 \\ 0 & -89.83 & -0.52 & 0 \\ 0 & 1027.45 & -10.89 & 0 \\ 0 & 0 & 1 & 0 \end{bmatrix}, \ B = \begin{bmatrix} 1 & 0 \\ 0 & 1 \\ 0 & 0 \\ 0 & 0 \end{bmatrix}.$$

For position control problem, the output equation is given by

$$y(t) = z_4(t). \tag{2.63}$$

Nonlinear surface and controller design: The control objective is to track a desired position as fast as possible in the presence of model uncertainty and external disturbances. In what follows next, we present a step by step approach to design a tracking controller using the proposed nonlinear sliding surface.

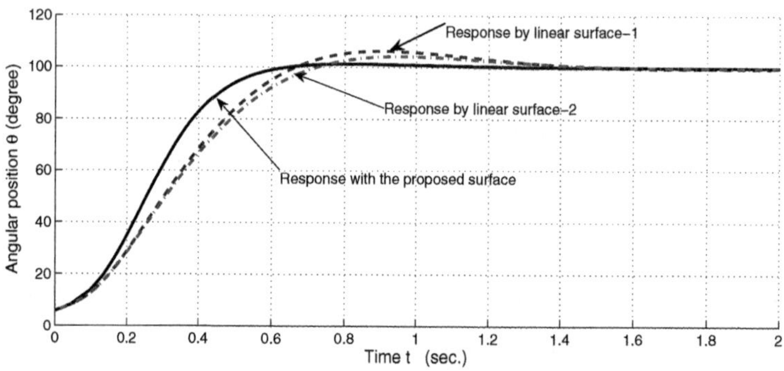

Fig. 2.4 Plot of angular position obtained from different linear surface and the proposed nonlinear sliding surface.

Step 1: Transform the system into regular form. Transformation matrix which transform the system into regular form can be computed as follows.

$$T_r = \begin{bmatrix} 0 & 0 & 1 & 0 \\ 0 & 0 & 0 & 1 \\ 1 & 0 & 0 & 0 \\ 0 & 1 & 0 & 0 \end{bmatrix}.$$

With this transformation matrix sub matrices of the system matrix in regular form can be obtained as follows:

$$A_{11} = \begin{bmatrix} -10.89 & 0 \\ 1 & 0 \end{bmatrix}, \ A_{12} = \begin{bmatrix} 0 & 1027.45 \\ 0 & 0 \end{bmatrix},$$

$$A_{21} = \begin{bmatrix} 0 & 0 \\ -0.52 & 0 \end{bmatrix}, \ A_{21} = \begin{bmatrix} -89.83 & 0 \\ 0 & -89.83 \end{bmatrix}, \ B_2 = \begin{bmatrix} 1 & 0 \\ 0 & 1 \end{bmatrix}.$$

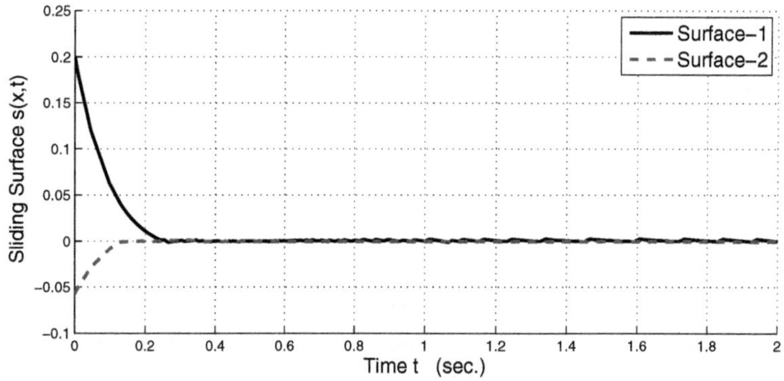

Fig. 2.5 Evolution of nonlinear sliding surface with time

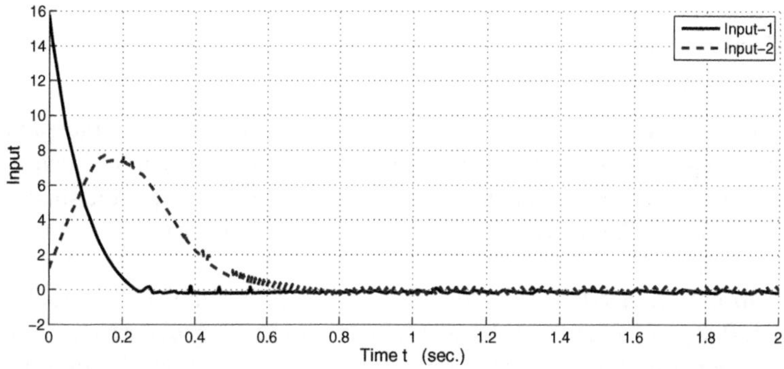

Fig. 2.6 Plot of inputs when nonlinear sliding surface is used with time

Step 2: Design F based on initial damping ratio ζ_1 and settling time t_{s1}. With $\zeta_1 = 0.5$ and $t_{s1} = 1.4\, sec.$, the matrix F can be computed by the pole placement technique as follows:

$$F = \begin{bmatrix} 0 & 0 \\ -0.0050 & 0.0318 \end{bmatrix}.$$

To achieve final damping ratio $\zeta_2 = 1$, solving (2.6) for $W = 10^{-4} \begin{bmatrix} 0.2000 & 0 \\ 0 & 0.2000 \end{bmatrix}$ gives matrix P as follows:

$$P = 10^{-4} \begin{bmatrix} 0.0180 & 0.0031 \\ 0.0031 & 0.6064 \end{bmatrix}.$$

The nonlinear function in (2.7) is implemented with $\beta_1 = 3.1$ and $\beta_2 = 3.2$. And, the initial condition is taken as $x_0 = \begin{bmatrix} 0.2 \; 0 \; 1 \; 0.1 \end{bmatrix}^T$. The initial value of output $y_0 = 0.1$ and the reference angular position to be tracked is 100 degree or 1.7453 rad. Therefore, in (2.33), it is selected to be $r = 1.7453$ rad.

Step 3: *Control law design*
To check robustness of the controller, the system is perturbed by a matched disturbance $\rho_{d2} = \begin{bmatrix} 0.2sin(t) & 0 \\ 0 & 0.2sin(t) \end{bmatrix}$. It should be noted that the disturbance is persistent. The control law in (2.39) is simulated with the following parameters

$$K = \begin{bmatrix} 10 & 0 \\ 0 & 10 \end{bmatrix}, Q = \begin{bmatrix} 0.21 & 0 \\ 0 & 0.21 \end{bmatrix}. \tag{2.64}$$

It is noted that diagonal entries of the matrix Q is just little more than the maximum value of disturbance. With these parameters and using parameters of nonlinear sliding surface discussed in Step 2, the control law in (2.39) can be computed.

To compare the performance of the proposed nonlinear sliding surface, two different linear surfaces are designed with different fixed damping ratios and settling times as follows.

- Linear surface-1 with $\zeta = 0.65$ and $t_s = 1.3$,
- Linear surface-2 with $\zeta = 0.7$ and $t_s = 1.2$.

While simulating the system with these linear surfaces, control law is the same as (2.39) with exception that nonlinear term becomes zero. The matrix F can be computed by the pole placement technique for each linear surface as per the surface parameters, i.e., ζ and t_s. Parameters K and Q are the same as in (2.64).

Analysis of simulation results: The stepper motor plant is simulated to track a step signal of 100 degree under the persistent influence of external disturbance. The controller is designed by the proposed nonlinear sliding surface and by two different linear surfaces. The response obtained by the proposed surface is compared with the responses obtained by linear surfaces. Fig. 2.4 shows the plot of angular position obtained when different surfaces are used. The same plot shows significant improvement in settling time without any overshoot. It can also be verified from Fig. 2.4 that peak overshoot and settling time can not be minimized for linear surface. It should be noted that the plant is subjected to continuous sinusoidal disturbances but effect of disturbances is not seen in the response. Fig. 2.5 shows evolution of nonlinear sliding surface with time. Plot of inputs is shown in Fig. 2.6. Chattering is observed in input plot because of sustaining disturbance is given to the system.

2.8 Conclusion

The design of nonlinear sliding surface has been proposed to improve the performance for different continuous-time uncertain systems. Existence of sliding mode has been proved. Position control examples for DC motor and stepper motor clearly show significant improvement of performance, compared to the conventional linear sliding surface. The proposed algorithm is able to achieve low overshoot and short settling time simultaneously for different uncertain continuous-time systems. Also, it turns out that robustness and high performance can be achieved with the proposed scheme. From the simulation results, we can observe that a significant reduction in settling time with negligible overshoot can be achieved with the proposed surface.

Chapter 3
High Performance Tracking Controller for Discrete Plant Using Nonlinear Surface

3.1 Introduction

In this chapter, a nonlinear surface is proposed to improve the performance of discrete-time system. Recall that, in the previous chapter a regulator design for continuous time system is considered. Due to the flexibility of implementation, most of the controllers are implemented through digital signal processor or high end microcontrollers. Due to this reason study and research on discrete sliding mode has received a considerable amount of attention (e.g., see [12, 10, 44, 8] and [54], among many). In this chapter, a step tracking control for general discrete-time multivariable system is considered based on DSMC with nonlinear sliding surface. To relax the need of measuring the entire state vector, the multi-rate output feedback (MROF) is used. Discrete-time system represented in delta operator is also analysed in this chapter to understand the effect of change in sampling time explicitly. This chapter reconstructs the research results proposed in [1, 3] as well as some additional results are included.

The brief outline of this chapter is as follows. Section 2 contains a brief review of the multirate output feedback strategy. The structure of nonlinear sliding surface and the proof of its stability is given in Section 3. Section 4 discusses two approaches to design control law, first is based on reaching law approach and the second is based on disturbance observer. Analysis of system represented with delta operator is presented in Section 3.5. Application and simulation results are presented in Section 6 followed by the conclusion in Section 7.

3.2 Multirate Output Feedback

Implementation of SMC law requires the availability of the entire state vector. But the complete state vector is seldom available. One of the ways to overcome this problem is to construct an observer. However, this may add additional complexity in the system. Moreover it is not always desirable to construct observer for an uncertain system. Hence one has to resort to output feedback design. It is well known

B. Bandyopadhyay et al.: Sliding Mode Control Using Novel Sliding Surfaces, LNCIS 392, pp. 41–64.
springerlink.com

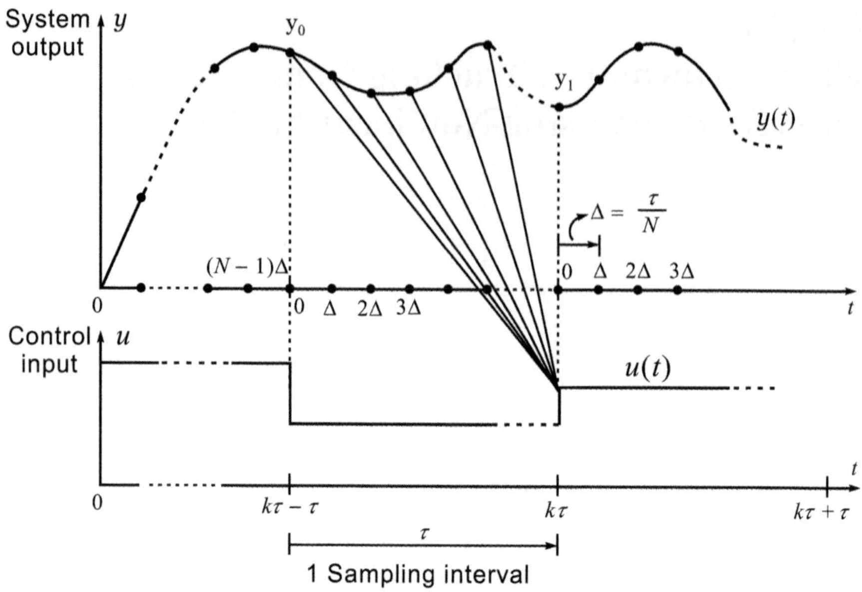

Fig. 3.1 Visualization of multirate sampling process

that a complete pole placement can not be achieved using *static* output feedback. A concept known as multirate output feedback technique [8, 54] which is of static output feedback kind and at the same time gives any arbitrary closed loop pole configuration. In what follows, we briefly review Multirate Output Feedback (MROF).

In MROF technique, the output is sampled at faster rate as compared to the control input. Consider the system described by the following equation

$$\dot{x}(t) = Ax(t) + Bu(t) + Bd(t) \tag{3.1a}$$
$$y(t) = C_1x(t). \tag{3.1b}$$

Let the above continuous system be sampled at τ period and under the assumption that disturbance does not change in relatively small sampling period. Discrete equivalent of the above continuous plant can be written as follows

$$x(k+1) = \Phi x(k) + \Gamma u(k) + \Gamma d(k), \tag{3.2a}$$
$$y(k) = C_1x(k). \tag{3.2b}$$

Where $x(k) \in \mathbb{R}^n$, $u(k) \in \mathbb{R}^m$, $y(k) \in \mathbb{R}^p$ are respectively the state, input, and controlled output of the system. Φ, Γ, C_1 are matrices of appropriate dimensions and $d(k)$ is matched uncertainty. The following assumptions are made:

A1. Disturbance remains constant during a sampling period τ.
A2. The pair (Φ, Γ) is stabilizable.
A3. The pair (Φ, C_1) is observable.

A1 is necessary because the matched perturbation is considered. It should be noted that without the assumption **A1** matched perturbation in continuous time becomes unmatched in discrete-time. Moreover, if the sampling rate is sufficiently high, then, it is reasonable to assume disturbance remains constant over a sampling period τ. Let the input u be applied with a sampling interval of τ seconds and the system output is sampled with a faster sampling period of $\Delta = \tau/N$ seconds, where N is an integer greater than or equal to the observability index [8] of the system. Let the triplet $(\Phi_\Delta, \Gamma_\Delta, C_1)$ represents the system in (3.1) sampled at Δ rate. Using the fact that u is unchanged in the interval $\tau < t < (k+1)\tau$, the τ system state dynamics can be constructed from the Δ system dynamics. Further, if the past N multirate-sampled system outputs are represented as

$$Y_k = \begin{bmatrix} y(k\tau - \tau) \\ y(k\tau - \tau + \Delta) \\ \vdots \\ y(k\tau - \Delta) \end{bmatrix}, \tag{3.3}$$

then τ system with multirate output samples can be represented as follows:

$$x(k+1) = \Phi x(k) + \Gamma u(k) + \Gamma d(k), \tag{3.4}$$

$$Y_{k+1} = C_0 x(k) + D_0 u(k) + D_0 d(k). \tag{3.5}$$

Where

$$C_0 = \begin{bmatrix} C_1 \\ C_1 \Phi_\Delta \\ \vdots \\ C_1 \Phi_\Delta^{N-1} \end{bmatrix}, D_0 = \begin{bmatrix} 0 \\ C_1 \Gamma_\Delta \\ \vdots \\ C_1 \Sigma_{j=0}^{N-2} \Phi_\Delta^j \Gamma_\Delta \end{bmatrix}. \tag{3.6}$$

From (3.4) and (3.5), $x(k)$ can be expressed using the past multirate output samples Y_k and the immediate past control input $u(k-1)$ as

$$x(k) = L_y Y_k + L_u u(k-1) + L_u d(k-1), \tag{3.7}$$

where

$$L_y = \Phi (C_0^T C_0)^{-1} C_0^T, \tag{3.8}$$

$$L_u = \Gamma - \Phi (C_0^T C_0)^{-1} C_0^T D_0. \tag{3.9}$$

It is clear from the above discussion that by the previous samples of output, immediate previous input and disturbance value, one can exactly compute the states of system. The visualization of multirate sampling is shown in Fig. 3.1.

3.3 Nonlinear Sliding Surface

This section discusses the design of sliding surface for general MIMO case with matched perturbation. Without loss of generality, the plant described by (3.2) can be transformed into regular form by using some orthogonal transformation matrix T_r as

$$z_1(k+1) = \Phi_{11}z_1(k) + \Phi_{12}z_2(k) \tag{3.10}$$
$$z_2(k+1) = \Phi_{21}z_1(k) + \Phi_{22}z_2(k) + \Gamma_2 u(k) + \tilde{d}(k) \tag{3.11}$$
$$y(k) = Cz(k). \tag{3.12}$$

Where $z_1 \in \mathbb{R}^{n-m}$, $z_2 \in \mathbb{R}^m$, $C = C_1(T_r)^{-1}$,
$\tilde{d}(k) = T_r\Gamma d(k) \in \mathbb{R}^m$, Γ_2 is a full rank $m \times m$ matrix, $z = T_r x = \begin{bmatrix} z_1 \\ z_2 \end{bmatrix}$. Define

$$\Phi_{reg} := \begin{bmatrix} \Phi_{11} & \Phi_{12} \\ \Phi_{21} & \Phi_{22} \end{bmatrix}. \tag{3.13}$$

Let a desired trajectory be $z_d := \begin{bmatrix} z_{1d} \\ z_{2d} \end{bmatrix}$.

$$c_1(k) := F - \Psi(y(k))\Phi_{12}^T P(\Phi_{11} - \Phi_{12}F), \tag{3.14a}$$
$$c^T := \begin{bmatrix} c_1(k) & I_m \end{bmatrix} \tag{3.14b}$$
$$e_1(k) := z_1(k) - z_{1d}(k) \tag{3.14c}$$
$$e_2(k) := z_2(k) - z_{2d}(k), \tag{3.14d}$$

where I_m is an identity matrix of $m \times m$, $\Psi(y(k))$, an $m \times m$ diagonal matrix with non-positive entries and F is chosen such that $(\Phi_{11} - \Phi_{12}F)$ has stable eigenvalues.

The sliding surface for the system in regular form is proposed as

$$\begin{aligned} s(k) &:= c^T(k)e(k), \\ &= \begin{bmatrix} c_1(k) & I_m \end{bmatrix} \begin{bmatrix} z_1(k) - z_{1d}(k) \\ z_2(k) - z_{2d}(k) \end{bmatrix}, \\ &= \begin{bmatrix} F - \Psi(y(k))\Phi_{12}^T P(\Phi_{11} - \Phi_{12}F) & I_m \end{bmatrix} \times \\ &\quad \begin{bmatrix} e_1(k) \\ e_2(k) \end{bmatrix}, \end{aligned} \tag{3.15}$$

In the above equation F is chosen such that $(\Phi_{11} - \Phi_{12}F)$ has stable eigenvalues and dominant poles are with low damping ratio. $\Psi(y(k))$, an $m \times m$ diagonal matrix with non-positive entries depending on the output, is used to change the damping ratio. P is an $(n-m) \times (n-m)$ positive definite matrix, obtained from the solution of the following Lyapunov equation

$$P = (\Phi_{11} - \Phi_{12}F)^T P(\Phi_{11} - \Phi_{12}F) + W, \tag{3.16}$$

for some positive definite matrix W. Such a P exists because $(\Phi_{11} - \Phi_{12}F)$ is a stable matrix.

Selection of Nonlinear Function $\Psi(y(k))$

The nonlinear function is used to change the system's closed loop damping ratio as the output approaches setpoint. It is chosen such that it changes from 0 to -β in a nonlinear way. One possible choice of $\Psi(y(k))$ is as follows

$$\Psi(y(k)) = \begin{bmatrix} \Psi(y(k))_1 & \dots & 0 \\ \vdots & . & \vdots \\ 0 & \dots & \Psi(y(k))_m \end{bmatrix}.$$

Where

$$\Psi(y(k))_i = -\beta_i \frac{\big| |y(k-1)_i - r(k)_i|^{\alpha_i} - |y(0)_i - r(0)_i|^{\alpha_i} \big|}{|y(0)_i - r(0)_i|^{\alpha_i}}, \quad i = 1,2....m. \tag{3.17}$$

In the above equation, $r(k)_i$ is a reference trajectory, β_i is used as a tuning parameter which contributes in deciding the final damping ratio and α_i decides the speed of change of damping ratio. Above choice of $\Psi(y(k))$ is similar in structure as suggested in [22] with reduced dimension. In the above equation at $k = 0$, output $y(k-1)$ can be approximated with $y(0)$. It should be noted that the choice of $\Psi(y(k))$ is not unique and any function with the above mentioned property can be used. Another possible choice [70] is as follows

$$\Psi(y(k))_i = -\beta_i e^{-\bar{k}_i |y(k-1)_i - r(k)_i|}, \tag{3.18}$$

where $\bar{k}_i$ is a positive constant. The nonlinear function should have ideally zero initial value. So initially damping ratio remains small which is contributed by F. Initial value of function given by (3.17) is zero but it needs more computation while (3.18) has some small non-zero initial value but from the implementation viewpoint it is simpler. $\Psi(y(k)_i)$ is chosen so that it satisfies the following condition

$$2\Psi(y(k)) + \Psi(y(k))\Phi_{12}^T P \Phi_{12} \Psi(y(k)) \leq 0. \tag{3.19}$$

Above condition is similar in structure as suggested in [22] with reduced size of matrices. During sliding mode $s(k) = 0$. So from (3.15),

$$e_2(k) = -c_1(k) e_1(k). \tag{3.20}$$

From (3.10) and (3.20) the system equation in sliding mode becomes

$$e_1(k+1) = \Phi_{11eq} e_1(k) + \phi_{12} \Psi(y(k)) \Phi_{12}^T P \Phi_{11eq} e_1(k) + \Phi_{11} z_{1d}(k) + \Phi_{12} z_{2d}(k) - z_{1d}(k+1), \tag{3.21}$$

where $\Phi_{11eq} = (\Phi_{11} - \Phi_{12}F)$. For tracking case, desired trajectory is consistently generated using system equation [78, 40] and due to this there exist some control $u_d(k)$ such that

$$\begin{aligned} z_{1d}(k+1) &= \Phi_{11}z_{1d}(k) + \Phi_{12}z_{2d}(k) \\ z_{2d}(k+1) &= \Phi_{21}z_{1d}(k) + \Phi_{22}z_{2d}(k) + \Gamma_2 u_d(k) \end{aligned} \tag{3.22}$$

Using (3.21) and (3.23), closed loop system (3.21) becomes

$$e_1(k+1) = \Phi_{11eq}e_1(k) + \phi_{12}\Psi(y(k))\Phi_{12}^T P\Phi_{11eq}e_1(k). \tag{3.23}$$

To prove the stability of the sliding surface, the stability of the above subsystem need to be proved which is proved in the following theorem. In further discussion, for the notational simplicity, the argument k for some variables is dropped i.e f stands for $f(k)$.

Theorem 3.3.1. *If $(\Phi_{11} - \Phi_{12}F)$ is stable and $\Psi(y(k))$ is defined by (3.17) or (3.18) which satisfy (3.19) then, subsystem in (3.23) is stable.*

Proof. Let a Lyapunov function for system (3.23) be defined as follows

$$V(k) = e_1^T(k)Pe_1(k).$$

Increment of $V(k)$ becomes

$$\begin{aligned} \Delta V(k) &= V(k+1) - V(k) \\ \Rightarrow \Delta V(k) &= e_1^T(k+1)Pe_1(k+1) - e_1^T(k)Pe_1(k) \\ &= \{\Phi_{11eq}e_1(k) + \Phi_{12}\Psi(y(k))\Phi_{12}^T P\Phi_{11eq}e_1(k)\}^T \times \\ &\quad P\{\Phi_{11eq}e_1(k) + \Phi_{12}\Psi(y(k))\Phi_{12}^T P\Phi_{11eq}e_1(k)\} - e_1(k)^T Pe_1(k) \\ &= e_1^T\Phi_{11eq}^T P\Phi_{12}\Psi(y(k))\Phi_{12}^T P\Phi_{11eq}e_1 + e_1^T\Phi_{11eq}^T e_1 - e_1^T Pe_1 + \\ &\quad e_1^T\Phi_{11eq}^T P\Phi_{12}\Psi(y(k))\Phi_{12}^T P\Phi_{11eq}e_1 + \\ &\quad e_1^T\Phi_{11eq}^T P\Phi_{12}\Phi_{12}^T\Phi_{12}\Psi(y(k))\Phi_{12}^T P\Phi_{11eq}e_1 \\ &= -e_1^T We_1 + e_1^T\Phi_{11eq}^T P\Phi_{12}\{2\Psi(y(k)) \\ &\quad +\Psi(y(k))\Phi_{12}^T P\Phi_{12}\Psi(y(k))\}\Phi_{12}^T P\Phi_{11eq}e_1 \end{aligned}$$

Let $e_1^T\Phi_{11eq}^T P\Phi_{12} = M^T$ therefore

$$\Delta V(k) = M^T\{2\Psi(y(k)) + \Psi(y(k))\Phi_{12}^T P\Phi_{12}\Psi(y(k))\}M - e_1^T We_1.$$

From the condition (3.19) it follows

$$\Delta V(k) \leq -e_1^T We_1.$$

So the system represented by (3.23) is stable. Hence the nonlinear surface is stable and thus the theorem is proved.

It should be noted that the system in sliding mode can be stabilized by any negative function $\Psi(y(k))$ which satisfies (3.19).

Remark 3.3.1. *During sliding, the system dynamics is decided by the subsystem described by (3.23). It can be observed that the poles of subsystem (3.23) changes as* $\Psi(y(k))_i$ *changes from 0 to* $-\beta_i$*. The main purpose of adding nonlinear part in sliding surface is to add significant value to control input as output reaches nearer to set point. The subsystem is stable for any non-positive value of* $\Psi(y(k)_i$ *where* $i = 1...m$ *which satisfy (3.19). Function* $\Psi(y(k))_i$ *changes from* 0 *to* $-\beta_i$ *as output tracks the reference signal. For any intermediate value of* $\Psi(y(k))_i$ *also the system (3.23) is stable. It is proved in [21] that introduction of this function changes damping ratio of system from initial value* ζ_1 *to final value* ζ_2 *where* $\zeta_2 > \zeta_1$*. Initially when* $\Psi(y(k)_i) = 0$ *damping ratio is determined by F which is designed for low damping ratio. When output reaches nearer to reference,* $\Psi(y(k)_i)$ *contributes significantly to damping ratio of the system and the steady state value of* $\Psi(y(k)_i)$ *becomes,* $\Psi(y(k))_i \approx -\beta_i$*. Therefore subsystem (3.23) can be written as*

$$e_1(k+1) = (\Phi_{11} - \Phi_{12}F - \Phi_{12}\beta\Phi_{12}^T P(\Phi_{11} - \Phi_{12}F))e_1(k), \tag{3.24}$$

where

$$\beta = \begin{bmatrix} \beta_1 & \dots & 0 \\ \vdots & \ddots & \vdots \\ 0 & \dots & \beta_m \end{bmatrix}.$$

Equation (3.24) decides the final damping ratio, therefore the parameter β *and matrix P should be designed so that the dominant poles of (3.24) have the desired damping ratio. To design matrix P and parameter* β *for the desired final damping ratio* (ζ_2)*, in [21, 22] zero placement approach is used.*

Another possible way to design P and β *is by trial and error. Choose diagonal matrix W and solve (3.16) for P. Through simulation, adjust the diagonal weight of W until satisfactory response is obtained. This technique generally gives better response after proper tuning. To design the matrix P, an LMI based formulation can also be used as it is proposed for continuous time system. While formulating LMI, condition (3.19) need to be satisfied to ensure stability along with the other constraints.*

3.4 Control Law

The sliding surface discussed in the previous section is nonlinear. Here surface parameters are changing at every instant. The control law should be chosen in such a way that from any initial condition, the system trajectory is attracted towards the sliding surface in finite time. To ensure the attractiveness of sliding surface condition $|s(k+1)| < |s(k)|$ should be satisfied. In discrete-time system, equivalent control [89, 76, 10] ensures attractiveness of sliding surface and keeps the trajectory on sliding surface at each sampling instant. However, for uncertain system to implement equivalent control law, the actual value of uncertainty is needed. In practice it is difficult to obtain the exact value of the uncertainty and this gives rise to a boundary layer around the sliding surface $s(k) = 0$. The width of the boundary layer depends on how the unknown disturbance is approximated. In this section two methods are presented for the design of controller. In the first method, a controller is designed based on the reaching law approach proposed in [82]. The second method approximates the current disturbance by its previous sampling instant disturbance value [110, 89]. The first method is simple from the implementation viewpoint and the width of the boundary layer is bounded by the spread of the disturbance. While the second method requires disturbance observer and the width of the boundary layer is bounded by the rate of change of the disturbance. Furthermore, it ensures almost complete rejection of slowly varying disturbances. In the following output feedback control law based on these two approaches is discussed.

3.4.1 Control Law Based on Reaching Law Approach

In this subsection a control law is derive based on reaching law approach which requires only disturbance bounds. Reaching law is so constructed that it replaces actual unknown disturbance terms in control law by its respective average values. From (3.15), $s(k+1)$ can be written as follows

$$
\begin{aligned}
s(k+1) &= c^T(k+1)T_r\{(x(k+1)) - x_d(k+1)\} \\
\Rightarrow s(k+1) &= c^T(k+1)T_r\Phi x(k) + c^T(k+1)T_r\Gamma u(k) + c^T(k+1)T_r\Gamma \\
&\quad d(k) - c^T(k+1)T_r x_d(k+1)
\end{aligned}
\tag{3.25}
$$

From (3.7) and (3.25) $s(k+1)$ can be written in terms of output as follows

$$
\begin{aligned}
s(k+1) = {} & c^T(k+1)T_r\Phi L_y Y_k + c^T(k+1)T_r\Phi L_u u(k-1) + \\
& c^T(k+1)T_r\Phi L_u d(k-1) + c^T(k+1)T_r\Gamma u(k) + \qquad (3.26) \\
& c^T(k+1)T_r\Gamma d(k) - \tilde{x}_d(k+1), \qquad (3.27)
\end{aligned}
$$

where

$$
\tilde{x}_d(k+1) = c^T(k+1)T_r x_d(k+1). \tag{3.28}
$$

To reach the sliding surface in one sampling instant reaching law becomes

$$
s(k+1) = 0.
$$

If control law is obtained from the above reaching law then it contains uncertain terms. However, in general only bounds of uncertainty are known, therefore as a remedy reaching law is constructed so that the actual unknown disturbance is replaced by its average value. In what follows, a modified reaching law [82] is used to obtain the control law.

Let us assume without loss of generality
$d_m(k) = c^T(k+1)T_r\Phi L_u d(k-1)$,
$d_n(k) = c^T(k+1)T_r\Gamma d(k)$.
Let us also assume that d_{mL}, d_{mU} and d_{nL}, d_{nU} are the lower and upper bounds of $d_m(k)$ and $d_n(k)$ respectively. Thus mean and spread of $d_m(k)$ and $d_n(k)$ are

$$\begin{aligned} d_{mM} &= \frac{d_{mL}+d_{mU}}{2}, \quad d_{mS} = \frac{d_{mU}-d_{mL}}{2} \\ d_{nM} &= \frac{d_{nU}+d_{nL}}{2}, \quad d_{nS} = \frac{d_{nU}-d_{nL}}{2}. \end{aligned} \tag{3.29}$$

In the above equation subscript M is for mean and subscript S for spread. Now consider the reaching law [82] for the uncertain system as follows

$$s(k+1) = d_m(k) - d_{mM} + d_n(k) - d_{nM} \tag{3.30}$$

With the above modification of reaching law, deviation of the trajectory from $s(k) = 0$ reduces in the presence of disturbance. Using the reaching law (3.30) and $s(k+1)$ from (3.27), the control law can be derived as follows

$$\begin{aligned} u(k) = & -(c^T(k+1)T_r\Gamma)^{-1}\{c^T(k+1)T_r\Phi L_y Y_k + \\ & c^T(k+1)T_r\Phi L_u u(k-1) + d_{mM} + d_{nM} - \tilde{x}_d(k+1)\}. \end{aligned} \tag{3.31}$$

From the reaching law (3.30), the magnitude of sliding mode band is given as follows

$$|s(k)| \le |d_{mS}| + |d_{nS}|. \tag{3.32}$$

With little abuse of notation, the modulus of s(k) represents modulus of each element of s(k) in subsequent analysis.

Control law (3.31) does not require the entire state vector. To implement the control law, controller uses past outputs, immediate past inputs and disturbance bounds as well as to evaluate $c^T(k+1)$ only $y(k)$ is needed. It should be noted that there is no switching term in the control law unlike used in [46]. For the system without disturbance, the trajectory with the proposed law does not deviate from the sliding surface.

3.4.1.1 Existence of Sliding Mode

With the proposed control law 3.31, boundary layer (the band of quasi sliding mode) for $s(k)$ is given in (3.32).

It can be shown that in the reaching phase i.e for $|s(k)| > |d_{mS}| + |d_{nS}|$ condition $|s(k+1)| < |s(k)|$ is satisfied.
Let Lyapunov function $v(k)$ be defined as $v(k) = |s(k)|$

$$\Delta v(k) = |s(k+1)| - |s(k)|$$

substituting $s(k+1)$ from (3.30)

$$\Delta v(k) = |d_m(k) - d_{mM} + d_n(k) - d_{nM}| - |s(k)|$$

$$\Rightarrow \Delta v(k) \leq |d_m(k) - d_{mM}| + |d_n(k) - d_{nM}| - |s(k)|$$

Further using (3.29)

$$\Delta v(k) \leq |d_{mS}| + |d_{nS}| - |s(k)|$$

So during the reaching phase i.e for $|s(k)| > |d_{mS}| + |d_{nS}|$.

$$\Delta v(k) < 0$$
$$\Rightarrow |s(k+1)| < |s(k)|$$

As seen from the above that during reaching phase when $|s(k)| > |d_{mS}| + |d_{nS}|$, reaching condition $|s(k+1)| < |s(k)|$ is satisfied by the control law (30), therefore the existence of DSM is proved. By considering bounded control input reaching condition is also proved in [10, 98] based on equivalent control approach for state feedback case. A similar approach can be used to prove the same result for output feedback case.

3.4.2 *Control Law with Disturbance Observer*

In the previous subsection control law is obtained from a reaching law in which the actual unknown disturbance is replaced by its average value. In this subsection the actual disturbance is approximated by the disturbance at the previous sampling instant. First the state vector of system and the previous instant disturbance are written in terms of previous outputs as given in [8, 55].

$$x(k) = \tilde{L}_y Y_k + \tilde{L}_u u(k-1) \tag{3.33}$$
$$d(k-1) = G_2 Y_k - G_2 D_0 u(k-1) \tag{3.34}$$

Where $\tilde{L}_y = \Phi G_1 + \Gamma G_2$ $\tilde{L}_u = \Gamma - (\Phi G_1 + \Gamma G_2) D_0$ While $\begin{bmatrix} G_1 \\ G_2 \end{bmatrix} = G = \begin{bmatrix} C_0 & D_0 \end{bmatrix}^{\dagger}$ is obtained by taking the generalized inverse (Moor-Penrose inverse) of $\begin{bmatrix} C_0 & D_0 \end{bmatrix}$ and Y_k, C_0, D_0 can be computed from (3.3) and (3.6).

To obtain the control law, reaching law $s(k+1) = 0$ is used. From (3.25) and reaching law $s(k+1) = 0$ one can write input $u(k)$ as follows

$$u(k) = -(c^T(k+1)T_r\Gamma)^{-1}\{c^T(k+1)T_r\Phi x(k) - c^T(k+1)T_r x_d(k+1) + d_n(k)\}, \tag{3.35}$$

where $d_n(k) = c^T(k+1)T_r\Gamma d(k)$ as defined earlier. In the above equation, except $d_n(k)$ everything is known. To obtain $d_n(k)$, one needs the actual disturbance $d(k)$ therefore as proposed in [110, 89], $d_n(k)$ is replaced with $d_n(k-1)$. To compute $d_n(k-1)$ one requires $d(k-1)$. As discussed earlier replacing $d_n(k)$ with the previous sampling instant disturbance $d_n(k-1)$ and substituting value of $x(k)$ from (3.33) control law can be written in terms of past output as follows

$$u(k) = -(c^T(k+1)T_r\Gamma)^{-1}\{c^T(k+1)T_r\Phi(\tilde{L}_yY_k + \tilde{L}_uu(k-1)) - \tilde{x}_d(k+1) + d_n(k-1)\}, \tag{3.36}$$

where $\tilde{x}_d(k+1)$ can be computed from (3.28) and

$$d_n(k-1) = c^T(k)T_r\Gamma(G_2Y_k - G_2D_0). \tag{3.37}$$

Using (3.25), (3.33), (3.36), (3.37) leads to

$$s(k+1) = d_n(k) - d_n(k-1)$$

From the above equation one can see that the sliding surface is bounded by the rate of change of disturbance. For constant and slowly varying disturbances this method gives excellent results. Taking into account the availability of high speed DSP and microcontrollers, the sampling time can be chosen sufficiently small and this leads to small value of boundary layer thickness $(d_n(k) - d_n(k-1))$.

3.5 Delta Operator Approach to Analyze Effect of Sampling Time

In the previous sections design of the nonlinear sliding surface is proposed in which system is represented by shift operator. Many system demand high sampling rates. System matrices become ill conditioned as sampling rate is increased when system is represented by shift operator. At high sampling, system represented in shift operator does not show any resemblance to the original continuous time system; the system matrix Φ approaches identity and the input matrix Γ approaches null matrix. To circumvent these problems it has been shown in [73, 74, 48] that when system dynamics is represented by delta operator, it leads superior numerical properties than corresponding shift operator does. The other motivation to use delta operator is that it allows unification of continuous and discrete-time results and smooth transition from either domain is possible. The use of high sampling rate also reduces the width of quasi sliding mode band and thus improves sliding accuracy. The other objective of this section is to understand effects of sampling time on design of nonlinear sliding surface. Consider the following continuous time linear uncertain system

$$\dot{x}(t) = Ax(t) + Bu(t) + Bd(t), \tag{3.38a}$$

$$y(t) = C_1x(t). \tag{3.38b}$$

In the above equation matrices A and B are of appropriate dimensions and $x \in \mathbb{R}^n$, $u \in \mathbb{R}$, $y \in \mathbb{R}$. Consider a case when input u remains constant over a sampling interval then the above system can be represented as described in [48] by the use of delta operator as follows

$$\delta x(k) = A_\delta x(k) + B_\delta u(k) + B_\delta d(k) \tag{3.39a}$$
$$y(k) = C_1 x(k) \tag{3.39b}$$

where

$$A_\delta = (I + \frac{A\tau}{2} + \frac{A^2\tau^2}{6} + ..)A \tag{3.40a}$$
$$B_\delta = (I + \frac{A\tau}{2} + \frac{A^2\tau^2}{6} + ..)B \tag{3.40b}$$

as mentioned earlier τ is sampling time.

Remark 3.5.1. *While deriving (3.39) from (3.38) it is assumed that disturbance remains constant during sampling interval τ. This assumption is made to ensure that matched uncertainty remains matched in discrete domain too. However, without this assumption it is possible to prove ultimate boundedness of motion. The main interest of this Section is to understand the discretized behavior of nonlinear sliding surface based algorithms therefore only matched uncertainty is considered. For the completeness, unmatched uncertainty is considered in the next chapter where performance improvement for time delay system is discussed.*

It is straight forward to convert the system in (3.39) to shift operator model given in (3.2) by using the following relation

$$\Phi = I + \tau A_\delta,$$
$$\Gamma = \tau B_\delta.$$

3.5.1 Sliding Surface Design

When B_δ matrix is of full rank then there exist a transformation matrix T_r such that system in (3.39) can be represented in regular form as follows

$$\delta\xi_1(k) = A_{\delta 11}\xi_1(k) + A_{\delta 12}\xi_2(k) \tag{3.41a}$$
$$\delta\xi_2(k) = A_{\delta 21}\xi_1(k) + A_{\delta 22}\xi_2(k) + B_{\delta 2}u(k) + B_\delta d(k) \tag{3.41b}$$
$$y(k) = C\xi(k) \tag{3.41c}$$

Where $\xi_1 \in \mathbb{R}^{n-1}$, $\xi_2 \in \mathbb{R}$, $C = C_1(T_r)^{-1}$, Nonlinear sliding surface is defined in δ coordinates as follows

$$s(k) := \left[\tilde{F} - \tilde{\Psi}(y(k))A_{\delta 12}^T\tilde{P}(I + \tau(A_{\delta 11} - A_{\delta 12}\tilde{F})\ 1\right]\begin{bmatrix}\xi_1(k)\\ \xi_2(k)\end{bmatrix}, \tag{3.42}$$

where $\tilde{\Psi}$ is a non positive function, $\tilde{F}$ is chosen such that eigen values of $(A_{\delta 11} - A_{\delta 12}\tilde{F})$ inside the circle of radius $1/\tau$ and center at$(-1/\tau, 0)$ in complex plane. This condition is needed to ensure the stability of system when dynamics represented in δ domain. The matrix $\tilde{P}$ is obtained by solving the following Lyapunov equation

$$(A_{\delta 11} - A_{\delta 12}\tilde{F})^T\tilde{P} + \tilde{P}(A_{\delta 11} - A_{\delta 12}\tilde{F}) + \tau(A_{\delta 11} - A_{\delta 12}\tilde{F})^T\tilde{P}(A_{\delta 11} - A_{\delta 12}\tilde{F}) = -Q. \tag{3.43}$$

It can be observed that the above equation reduces to standard Lyapunov equation in continuous time domain as τ approaches zero. Function $\tilde{\Psi}$ can be defined in similar way as discussed earlier, one possible way to select as follows

$$\tilde{\Psi} = \tilde{\beta} e^{-\tilde{k}|y(k)|}, \tag{3.44}$$

where $\tilde{\beta}$ and $\tilde{k}$ are positive constants. There are many other ways to construct such a function. The chosen function should be negative and its absolute value should increase from a small initial value to a finite final value in synchronization with output.

In SMC, system trajectory is forced to be on sliding surface from any initial condition in finite time by some control law therefore stability of surface is the first requirement. Before we proceed the proof of stability, for notational simplicity let us define:

$$c_{1\delta}(k) := \tilde{F} - \tilde{\Psi}(y(k))A_{\delta 12}^T\tilde{P}(I + \tau(A_{\delta 11} - A_{\delta 12}\tilde{F})) \tag{3.45}$$

$$c_{\delta}(k) := \left[\tilde{F} - \tilde{\Psi}(y(k))A_{\delta 12}^T\tilde{P}(I + \tau(A_{\delta 11} - A_{\delta 12}\tilde{F}))\ 1\right] \tag{3.46}$$

$$:= \left[c_{1\delta}(k)\ 1\right] \tag{3.47}$$

During ideal sliding mode $s(k) = 0$, therefore

$$\begin{aligned} \delta\xi_1 &= A_{\delta 11}\xi_1 - A_{\delta 12}c_{1\delta}(k)\xi_1, \\ &= A_e\xi_1 + M_1\xi_1, \end{aligned} \tag{3.48}$$

where $A_e = A_{\delta 11} - A_{\delta 12}\tilde{F}$,
$M_1 = \tilde{\Psi}(y(k))A_{\delta 12}A_{\delta 12}^T\tilde{P}(I + \tau A_e)$.
To ensure stability of sliding surface in (3.42) we need to prove stability of (3.48).

Remark 3.5.2. *When a matrix is termed stable in delta domain that indicates that all eigen values are in a circle of radius* $1/\tau$ *and center at* $(-1/\tau, 0)$ *and henceforth a stable matrix in delta domain should be interpreted in this way.*

Theorem 3.5.1. *System in (3.48) is stable with* $\tilde{\Psi}(y(k))$ *is defined by (3.44) and*

$$0 \le -\tilde{\Psi}(y(k)) \le \frac{2}{\tau\|A_{\delta 12}^T P A_{\delta 12}\|} \tag{3.49}$$

Proof. Let Lyapunov function for the system in (3.48) be

$$
\begin{aligned}
V(k) &= \xi_1^T \tilde{P}\xi \\
\Rightarrow \delta V(k) &= \frac{V(k+1)-V(k)}{\tau} \\
&= \frac{\xi_1^T(k+1)\tilde{P}\xi(k+1)-\xi_1^T\tilde{P}\xi}{\tau} \\
&= \frac{[\xi_1(K+1)-\xi_1]^T\tilde{P}[\xi_1(K+1)-\xi_1]+2\xi_1^T(k+1)\tilde{P}\xi_1-2\xi_1^T\tilde{P}\xi}{\tau} \\
&= \delta\xi_1^T\tilde{P}\tau\delta\xi_1+2(\frac{\xi_1(k+1)-\xi_1(k)}{\tau})^T\tilde{P}\xi_1
\end{aligned}
$$

$$
\begin{aligned}
&= \delta\xi_1^T\tilde{P}\tau\delta\xi_1+2\delta\xi_1^T\tilde{P}\xi_1 \\
&= [A_e\xi_1+M_1\xi_1]^T\tilde{P}\tau[A_e\xi_1+M_1\xi_1]+2[A_e\xi_1+M_1\xi_1]^T\tilde{P}\xi_1 \\
&= \xi_1^TA_e^T\tilde{P}\tau A_e\xi_1+\xi_1^TM_1^T\tilde{P}\tau A_e\xi_1+\xi_1^TM_1^T\tilde{P}\tau M_1\xi_1+ \\
&\quad \xi_1^TA_e^TP\tau M_1\xi_1+2\xi_1^TA_e^T\tilde{P}\xi_1+2\xi_1^TM_1\tilde{P}\xi_1 \\
&= \xi_1^T(A_e^T\tilde{P}\tau A_e+A_e^TP+PA_e)\xi_1+2\xi_1^TM_1^T\tilde{P}\tau A_e\xi_1+ \\
&\quad \xi_1^TM_1^T\tilde{P}\tau M_1\xi_1+2\xi_1^TM_1^T\tilde{P}\xi_1 \\
&= -\xi_1^TQ\xi_1+2\xi_1^TM_1^T\tilde{P}(A_e\tau+I)\xi_1+\xi_1^TM_1^T\tilde{P}\tau M_1\xi_1 \\
&\quad -\xi_1^TQ\xi_1+2\Psi\xi_1^T(A_e\tau+I)^T\tilde{P}A_{\delta 12}A_{\delta 12}^T\tilde{P}(A_e\tau+I)\xi_1+ \\
&\quad \Psi^2\tau\xi_1^T(A_e\tau+I)^T\tilde{P}A_{\delta 12}A_{\delta 12}^T\tilde{P}A_{\delta 12}A_{\delta 12}^T\tilde{P}(I+\tau A_e)\xi_1 \\
&= -\xi_1^TQ\xi_1+\xi_1^T((A_e\tau+I)^T\tilde{P}A_{\delta 12})\{2\Psi I+\Psi^2\tau(A_{\delta 12}^T\tilde{P}A_{\delta 12})\}(A_{\delta 12}^T\tilde{P}(I+\tau A_e))\xi_1
\end{aligned}
$$

Using (3.49) it follows

$$\delta V(k) < 0.$$

Thus nonlinear surface in (3.42)is stable and it completes the proof.

Remark 3.5.3. *It can be easily observed that as τ approaches zero then any value of Ψ stabilizes the system like continuous time case. However, with finite sampling time Ψ is to be restricted by the condition (3.49). It should also be noted that a small sampling time allows large value of Ψ. The main advantage of delta operator is it allows arbitrarily small sampling rate unlike shift operator case where very small sampling rate results in ill conditioning of system matrices.*

3.5.2 Existence Condition

In the previous section we proved stability during sliding mode. In this subsection a control law is proposed so that from any initial condition state trajectory is forced on sliding surface (or in a band around it)and then after it remains in this band. It is a well established fact that with finite switching frequency it is not possible to ensure ideal sliding mode. With finite switching frequency, switching function $s(k)$ remains in a band around $s(k) = 0$. In this section we also analyze the effect of sampling time on this band. Before we go for the proof of existence of sliding mode, we evaluate $\delta s(k)$.

$$\begin{aligned}
\delta s(k) &= \frac{s(k+1) - s(k)}{\tau} \\
&= \frac{c_\delta(k+1)\xi(k+1) - c_\delta(k)\xi(k)}{\tau} \\
&= \frac{c_\delta(k+1)(\xi(k+1) - \xi(k))}{\tau} + \frac{\xi(k)(c_\delta(k+1) - c_\delta(k))}{\tau} \\
&= c_\delta(k+1)\delta\xi(k) + \xi(k)\delta c_\delta(k) \qquad (3.50)
\end{aligned}$$

Define

$$A_{\delta r} := \begin{bmatrix} A_{\delta 11} & A_{\delta 12} \\ A_{\delta 21} & A_{\delta 22} \end{bmatrix}, \qquad (3.51)$$

$$B_{\delta r} := \begin{bmatrix} 0 \\ B_{\delta 2} \end{bmatrix}, \qquad (3.52)$$

$$\xi(k) := \begin{bmatrix} \xi_1(k) \\ \xi_2(k) \end{bmatrix}. \qquad (3.53)$$

By using the above definitions system equation in (3.41)

$$\delta\xi(k) = A_{\delta r}\xi + B_{\delta r}u(k) + B_{\delta r}d(k) \qquad (3.54)$$

A control law which ensures the existence condition is proved in the following theorem.

Theorem 3.5.2. *Control law*

$$u(k) = -B_{\delta 2}^{-1}(c_\delta(k+1)A_{\delta r}\xi(k) + \xi(k)\delta c_\delta(k) - Qsgn(s)), \qquad (3.55)$$

where

$$Q \geq \|d(k)_{max}\|, \qquad (3.56)$$

enforces trajectory of system (3.41) from any initial condition to a band equals to $2\tau Q$ *around* $s(k) = 0$.

Proof. Let a Lyapunov function be

$$\begin{aligned}
V_1(k) &= |s(k)| \\
\Rightarrow \delta V_1(k) &= \frac{V_1(k+1) - V_1(k)}{\tau} \\
&= \frac{|s(k+1)| - |s(k)|}{\tau} \qquad (3.57)
\end{aligned}$$

(3.58)

Using (3.50), (3.54), and control law (3.55), $\delta s(k)$ becomes

$$\begin{aligned}\delta s(k) &= d(k) - Qsgn(s(k)),\\ \frac{s(k+1)-s(k)}{\tau} &= d(k) - Qsgn(s(k)),\\ s(k+1) &= s(k) + \tau d(k) - Q\tau sgn(s(k)). \end{aligned} \tag{3.59}$$

By using condition (3.56), it is easy to verify that the ultimate bound of $s(k)$

$$|s(k)| \leq 2Q\tau. \tag{3.60}$$

It is worth noting that the above bound of $s(k)$ will be reached in finite time. From the above condition, it follows that

$$|s(k)| > 2Q\tau \Rightarrow |s(k+1)| - |s(k)| \leq 0. \tag{3.61}$$

By use of the above condition, (3.57) it follows

$$\text{when } |s(k)| > 2Q\tau \Rightarrow V_1(k) < 0. \tag{3.62}$$

This completes the proof it has been proved that control law (3.55) ensures that $s(k)$ decreases when it is outside the ultimate band $2Q\tau$.

3.6 Magnetic Tape Position Tracking

In this section a magnetic-tape-drive servo is presented to illustrate the proposed method. The control of tension and position of a moving tape is a generic control problem in industries. Applications vary widely from digital tape transport to thin film manufacturing. For position control of read-write head, the detailed dynamics

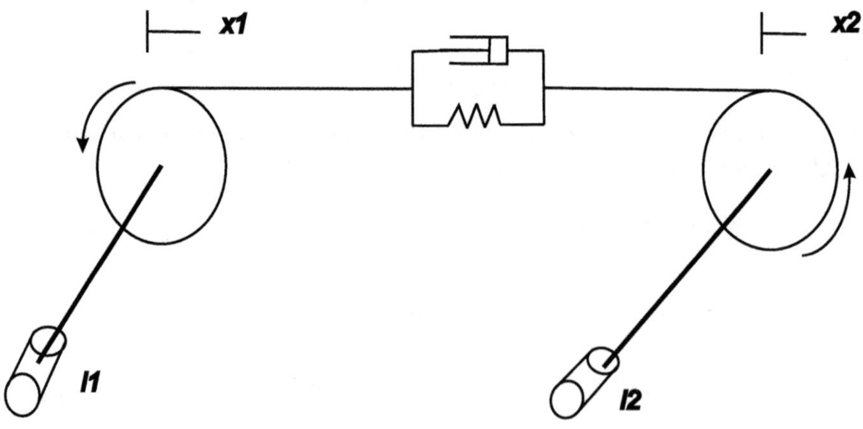

Fig. 3.2 Schematic diagram of magnetic tape position control

are given in [42]. Here the control system requirement is to achieve the commanded position of tape over read-write head. While achieving the required position, the control system should maintain specific tension in the tape. All the simulations are made with a continuous plant model and control is applied through a zero order hold.

The continuous time model of the system is given as

$$\dot{x} = Ax + Bu + Bd,$$
$$y = C_1 x,$$

where

$$A = \begin{bmatrix} 0 & 0 & -10 & 0 \\ 0 & 0 & 0 & 10 \\ 3.3150 & -3.3150 & -0.5882 & -0.5882 \\ 3.3150 & -3.3150 & -0.5882 & -0.5882 \end{bmatrix},$$

$$B = \begin{bmatrix} 0 & 0 \\ 0 & 0 \\ 8.5330 & 0 \\ 0 & 8.5330 \end{bmatrix},$$

$$C_1 = \begin{bmatrix} 0.5 & 0.5 & 0 & 0 \\ -2.113 & 2.113 & 0.375 & 0.375 \end{bmatrix}.$$

Here $x = [x_1 \ \ x_2 \ \ \omega_1 \ \ \omega_2]^T$ x_1, x_2 are the positions of the tape at capstans (in mm) and ω_1, ω_2 are angular velocities motor/capstan assemblies; $u = [i_1 \ \ i_2]^T$ i_1 and i_2 are currents supplied to drive motors and $d(t)$ is smooth disturbance. The output of the system is given by

$$y(t) = \begin{bmatrix} y_1(t) \\ y_2(t) \end{bmatrix} = \begin{bmatrix} f_p(t) \\ T_e(t) \end{bmatrix},$$

where f_p is the position of the tape over read-write head in mm, and T_e is the tension in the tape in N. The following control objectives need to be achieved.

1. Magnetic tape should achieve commanded position over read-write head.
2. Settling time t_s should be less than 2.5 seconds
3. Overshoot should be less than 20 %.
4. The tape tension, T_e should be 2N with the constraint $0 < T_e < 4N$.
5. Input current should not exceed 1A at each drive motor.
6. The controlled system should be robust.

To design a discrete sliding mode controller, discretize the model with sampling rate $\tau = 0.05$ as suggested in [42]. Discrete model of the plant under assumption A_1 is given as follows

$$x(k+1) = \Phi x(k) + \Gamma u(k) + \Gamma d(k), \tag{3.63}$$
$$y(k) = C_1 x(k), \tag{3.64}$$

where

$$\Phi = \begin{bmatrix} 0.9599 & 0.0401 & -0.4861 & 0.0139 \\ 0.0401 & 0.9599 & -0.0139 & 0.4861 \\ 0.1566 & -0.1566 & 0.9321 & -0.0679 \\ 0.1566 & -0.1566 & -0.0679 & 0.9321 \end{bmatrix},$$

$$\Gamma = \begin{bmatrix} -0.1049 & 0.0017 \\ -0.0017 & 0.1049 \\ 0.4148 & -0.0118 \\ -0.0118 & 0.4148 \end{bmatrix}.$$

$$C_1 = \begin{bmatrix} 0.5000 & 0.5000 & 0 & 0 \\ -2.1130 & 2.1130 & 0.3750 & 0.3750 \end{bmatrix}. \tag{3.65}$$

First consider $d(k) = 0$ to show improvement with the proposed sliding surface then disturbance will added to show robustness of the controller.

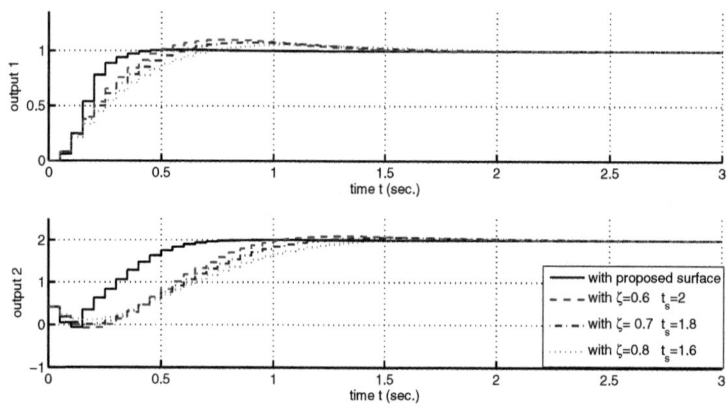

Fig. 3.3 Response of output $y(k)$ with different sliding surfaces

Step 1: Transform the system in regular form by an appropriate T_r matrix. Let $z = T_r x$ therefore in regular form system equation becomes

$$z_1(k+1) = \Phi_{11} z_1 + \Phi_{12} z_2, \tag{3.66}$$

$$z_2(k+1) = \Phi_{21} z_1 + \Phi_{22} z_2 + \Gamma_2 u + \tilde{d}(k), \tag{3.67}$$

$$y = Cz. \tag{3.68}$$

where $\Phi_{11} = \begin{bmatrix} 0.8802 & -0.0021 \\ -0.0021 & 0.8803 \end{bmatrix}$, $\Phi_{12} = \begin{bmatrix} -0.4727 & 0.0237 \\ 0.0185 & 0.4728 \end{bmatrix}$, $\Phi_{21} = \begin{bmatrix} 0.1678 & 0.1325 \\ -0.1476 & -0.1709 \end{bmatrix}$, $\Phi_{22} = \begin{bmatrix} 1.0177 & 0.1057 \\ 0.1057 & 1.0059 \end{bmatrix}$, $\Gamma_2 = \begin{bmatrix} 0.4281 & -0.0238 \\ -0.0000 & -0.4274 \end{bmatrix}$.
$C = \begin{bmatrix} 0.4768 & -0.4932 & -0.1246 & -0.1179 \\ -1.9866 & -1.9208 & 0.8623 & -0.9116 \end{bmatrix}$

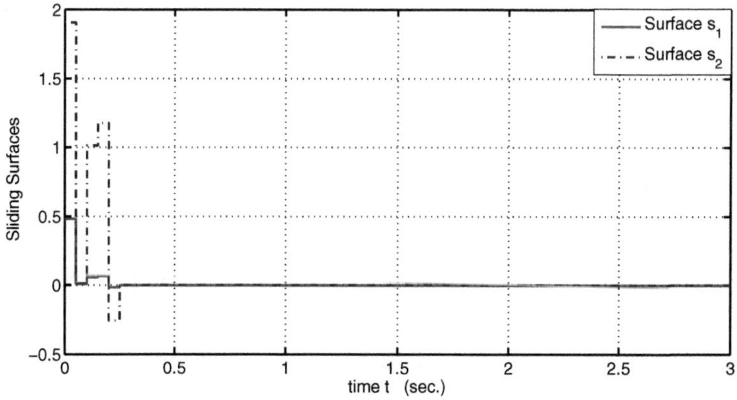

Fig. 3.4 Plot of nonlinear sliding surfaces with time.

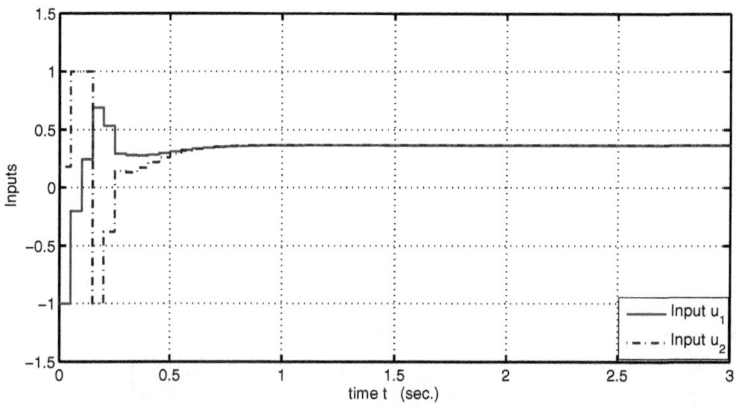

Fig. 3.5 Plot of input when nonlinear sliding surface is used

Based on output requirement $y_d = [1 \quad 2]^T$ required constant state trajectory is $x_d = [0.5267 \;\; 1.4733 \;\; 0 \;\; 0]^T$ and $r(k) = [1 \;\; 2]^T$

Step 2: Design of nonlinear sliding surface.
Nonlinear sliding surface is composed of a constant and a nonlinear term. Initially the nonlinear term is zero, therefore the constant term decides initial damping ratio (ζ_1) and settling time. For initial settling time $t_{s1} = 1.3sec.$ and initial damping ratio $\zeta_1 = 0.7054$ matrix F can be computed as

$$F = \begin{bmatrix} -0.0847 & -0.2713 \\ -0.2809 & 0.0815 \end{bmatrix}.$$

For $W = \begin{bmatrix} 3.0336 & -0.0000 \\ -0.0000 & 3.0336 \end{bmatrix}$ solving Lyapunov equation (3.16) for P gives

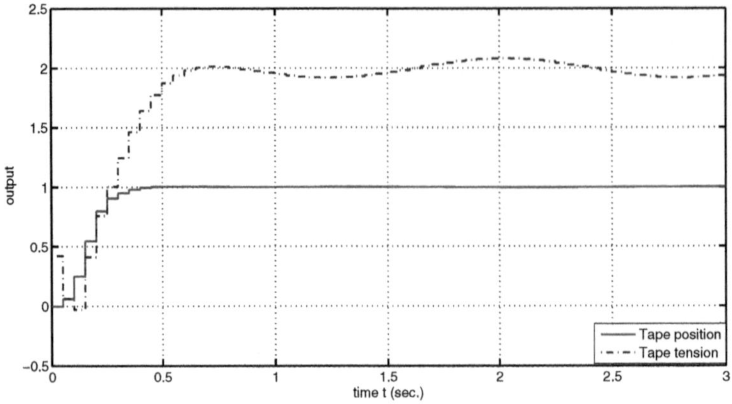

Fig. 3.6 Plot of outputs, when nonlinear sliding surface is used with plant disturbance and control is based on reaching law approach.

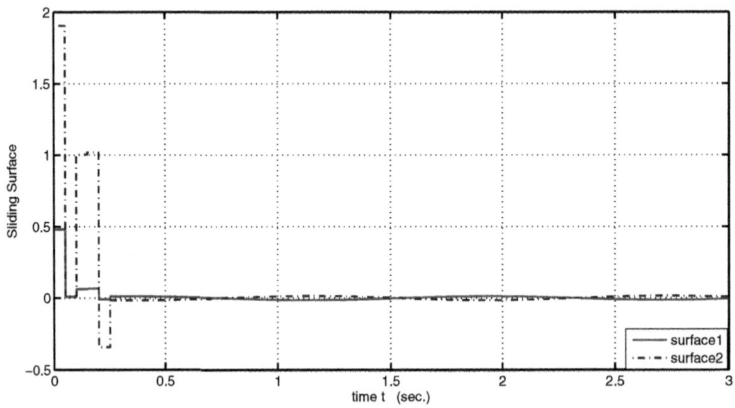

Fig. 3.7 Plot of nonlinear sliding surface with disturbance and control is based on reaching law approach.

$$P = \begin{bmatrix} 3.7682 & 0 \\ 0 & 3.7682 \end{bmatrix}.$$

Matrix P is positive definite as required. Matrix $\Psi(y(k))$ is given as follows $\Psi(y(k)) = \begin{bmatrix} \Psi(y(k))_1 & 0 \\ 0 & \Psi(y(k))_2 \end{bmatrix}$. Functions $\Psi(y(k))_1$ and $\Psi(y(k))_2$ can be computed from (3.17) with the following parameters $\beta_1 = -0.0039$, $\alpha_1 = 5$, $\beta_2 = -0.0039$, $\alpha_2 = 5$. From the above values switching function $s(k)$ can be computed.

Step 3: Design of control law.
From (3.31) control law with $d(k) = 0$

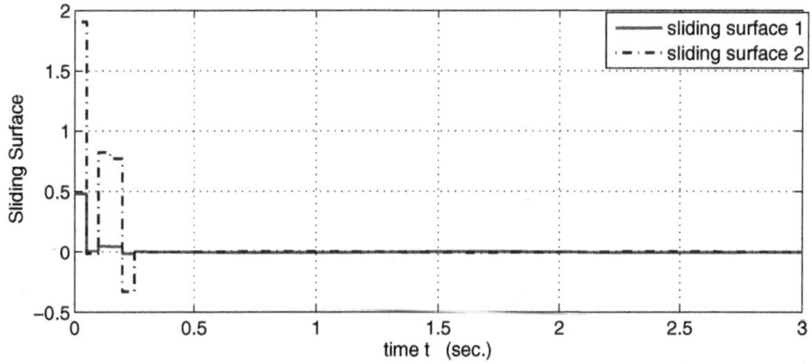

Fig. 3.8 Plot of outputs, when nonlinear sliding surface is used with plant disturbance and control law with disturbance observer.

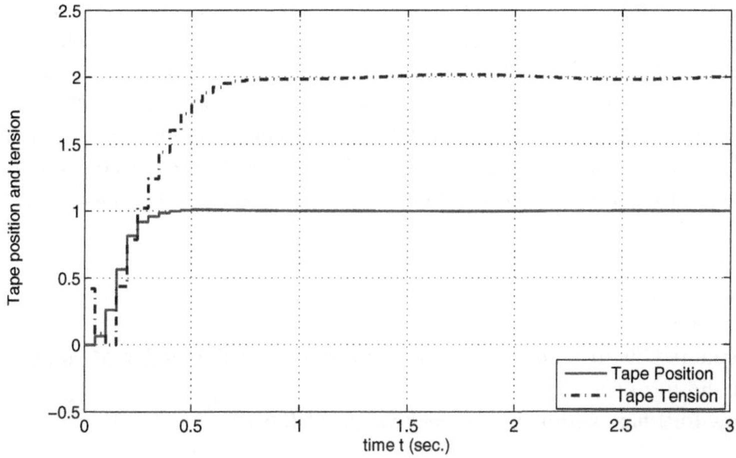

Fig. 3.9 Plot of nonlinear sliding surface with disturbance and control law with disturbance observer.

$$
\begin{aligned}
u(k) = & -(c^T T_r \Gamma)^{-1}\{c^T(k+1)T_r \Phi L_y Y_k + \\
& c^T(k+1)T_r \Phi L_u u(k-1) - \tilde{x}_d(k+1)\}.
\end{aligned} \tag{3.69}
$$

In the above equation $c^T(k+1)$ is computed from y(k). For the observability index 4, N is chosen 4. Let $L_y = [L_1 \ : \ L_2]$

$$
L_1 = \begin{bmatrix} -0.5000 & 0.0178 & 0.0000 & -0.0339 \\ -0.5000 & -0.0178 & 0.0000 & 0.0339 \\ 2.4000 & -0.5594 & 0.8000 & -0.1870 \\ -2.4000 & -0.5594 & -0.8000 & -0.1870 \end{bmatrix}
$$

$$L_2 = \begin{bmatrix} 0.5000 & -0.0845 & 1.0000 & -0.1336 \\ 0.5000 & 0.0845 & 1.0000 & 0.1336 \\ -0.8000 & 0.1819 & -2.4000 & 0.5436 \\ 0.8000 & 0.1819 & 2.4000 & 0.5436 \end{bmatrix}$$

$$L_u = \begin{bmatrix} -0.0231 & 0.0102 \\ -0.0102 & 0.0231 \\ 0.1903 & -0.0764 \\ -0.0764 & 0.1903 \end{bmatrix}$$

$\tilde{x}_d(k+1)$ can be computed from (3.28).

Design of linear sliding surface. Performance of the proposed nonlinear sliding surface is compared with the controller designed with different linear sliding surfaces. The system output response is plotted for different sliding surfaces and the proposed sliding surface. Parameters of linear sliding surface(ζ and t_s) are chosen based on the values of these parameters taken at different instants when the nonlinear sliding surface is used. When the nonlinear sliding surface is used, the poles of the closed loop system changes as output approaches the reference. This changes the damping ratio and settling time from its initial values ζ_1 and t_{s1} to final values ζ_2 and t_{s2}. During the course of change, at different instants, the system has different damping ratios and settling times. Following three different linear sliding surfaces are designed
Linear surface-1 with $\zeta = 0.6, t_s = 2.0$
Linear surface-2 with $\zeta = 0.7, t_s = 1.8$
Linear surface-3 with $\zeta = 0.8, t_s = 1.6$
Responses obtained by different sliding surfaces are compared with that of obtained by nonlinear sliding surface.

Control law for linear sliding surface is given as follows

$$\begin{aligned} u(k) = & -(\bar{c}^T T_r \Gamma)^{-1}\{\bar{c}^T T_r \Phi L_y Y_k + \bar{c}^T T_r \Phi L_u u(k-1) + \\ & d_{m0} + d_{10} - \tilde{x}_d(k+1)\}. \end{aligned} \tag{3.70}$$

where

$$\tilde{x}_d(k+1) = \bar{c}^T T_r x_d(k+1).$$

In above control law $\bar{c}^T$ can be designed to obtain desired damping ratio by using regular form and pole placement approach.

3.6.1 Comparison with Different Linear Sliding Surfaces

Responses of tape position $y_1(k)$ and tension in tape $y_2(k)$ are plotted with nonlinear sliding surface and different linear sliding surfaces. Fig. 3.3 shows the response of $y(k)$ with the controller designed with different sliding surfaces. The plot clearly shows that with the nonlinear sliding surface performance improves significantly. With the proposed surface position settles in 0.5 seconds without any overshoot.

Table 3.1 Settling time of tape position obtained by the proposed sliding surface versus surfaces with diff. ζ

Type of sliding surface	Peak (%) overshoot	Settling t_s(sec.) Time
surface with ζ=0.6 and $t_s = 2$	10 %	2.4
surface with ζ=0.7 and $t_s = 1.8$	7.5%	1.5
surface with ζ=0.8 and $t_s = 1.6$	5.4%	1.7
Proposed surface	1%	0.5

Minimum improvement in settling time for position is 50%. It can be seen that the proposed surface ensures quick response without any overshoot. Switching function $s(k)$ goes to zero in five sampling instants which can be seen from Fig. 3.4. Plot of input is shown in Fig. 3.5 which confirms that chattering is eliminated because of equivalent control.

3.6.2 *Nonlinear Sliding Surface with Disturbance*

Plant is perturbed by external matched disturbance $d(k) = 0.04sin(8\pi k\tau)$ to validate robustness property of the proposed surface. The responses obtained by the control law designed based on reaching law approach (3.31) and the control law with disturbance observer (3.36) are compared.

3.6.2.1 Control Law Designed Based on Reaching Law Approach

Response of the outputs and sliding surfaces are plotted in Figs. 3.6 and 3.7. The control law in (3.31) is implemented. In symmetric disturbance upper and lower bounds are same, therefore $d_{mM} = d_{nM} = 0$ Amplitude of sliding mode band is

$$|d_{mS}| + |d_{nS}| = \begin{bmatrix} 0.0545 \\ 0.0559 \end{bmatrix}$$

In the above, the first entry corresponds to the first sliding surface band size and the second is the band size for the second sliding surface.

3.6.2.2 Control Law Designed Based on Disturbance Observer

Following the procedure described in Section-IV control law as per (3.36) is designed. To implement the control law constant matrices $\tilde{L}_y$,$\tilde{L}_u$,G_1,G_2 are obtained as discussed in Section-IV. Fig. 3.8 shows plot of tape position and tension in the tape when the plant is subjected to a disturbance. It can be seen that the effect of disturbance is very small on both outputs. Fig. 3.9 shows evolution of switching function. It can be verified that the band is significantly reduced because of the disturbance observer.

3.7 Conclusion

The design of a nonlinear sliding surface, which allows the closed-loop system to simultaneously achieve low overshoot and low settling time, has been presented in this chapter. It has been shown how this high performance can be combined with high robustness (to matched uncertainty) and effective disturbance rejection. By application, it has been shown (Table I) that the nonlinear surface gives a minimum of 50% improvement in settling time (and negligible overshoot) when compared to linear surfaces. In the following chapter we will discuss the performance improvement of a system with the additional complexities like time-delay and the presence of unmatched perturbations.

Chapter 4
An Improvement in Performance of Input-Delay System Using Nonlinear Sliding Surface

4.1 Introduction

It is well recognized that the existence of a time delay may affect the performance or result in loss of stability as shown in [65]. Recently many methods have been published in [106, 104, 57, 83, 103] on the design of control laws for time-delay systems; see the references therein. In [77], it has been shown that all the proposed methods for time delay systems use prediction of state either explicitly or implicitly. Control algorithms based on state prediction were first proposed in [72]. Furukawa and Shimemura [43] proposed a predictor-observer based scheme to control plants with time-delay. Using the predictor, the original input-delay system can be converted into a delay-free system and the problem reduces to finite dimensions. The predictor is used when the delay is known which restricts its scope. However, Lozano *et al.* [71] consider uncertainty in the knowledge of delay with a state predictor based scheme. Recently, in [104], sliding mode control is proposed based on a discrete predictor for a regulator case. In [71], a state predictor based state feedback control law is proposed; and the authors also propose a predictor in the discrete-time framework. The performance of a system is adversely affected by a delay in the input as shown in [65, 83], which necessitates compensation. In this chapter, it has been shown that how the performance of input-delay systems can be improved by using a nonlinear sliding surface unlike the use of a linear sliding surface (linear in the predicted states). A nonlinear sliding surface is designed in predicted state. Furthermore, it has been shown if performance of the system transformed in the predicted state is improved then it leads to the improvement of the performance of the original time-delay system. The general uncertain system is considered which contains both matched and unmatched perturbations. It is an established fact that for an uncertain system, discrete-time sliding mode is possible only in the vicinity of sliding surface $s(k) = 0$. To ensure ideal sliding motion $s(k) = 0$ for an uncertain system, the exact value of disturbance/uncertainty is needed. In general, the exact value of disturbance/uncertainty is not known. Therefore, in this chapter an ultimate boundedness of resulting motion is proved. This chapter extends the results of the previous chapter for a system with a delay and having both matched and unmatched uncertainties. This chapter is based on authors work in [7].

B. Bandyopadhyay et al.: Sliding Mode Control Using Novel Sliding Surfaces, LNCIS 392, pp. 65–81.
springerlink.com

A brief outline of the chapter is as follows. The discrete-time state predictor is given Section 2. Section 3 contains the structure of the nonlinear sliding surface. Section 4 presents the control law. Stability of motion and ultimate boundedness is proved in Section 5. An example and simulation results are presented in Section 6 followed by the conclusions in Section 7.

4.2 Discrete-Time Predictor

In this section, we review the concept of the discrete-time state predictor as proposed in [71, 104] for the system

$$\begin{aligned} x(k+1) &= \Phi x(k) + \Gamma u(k-h) + D\rho(k). \\ y(k) &= C_1 x(k). \\ u(k) &= \Theta(k) \quad k = -h, -h+1, ...0 \end{aligned} \tag{4.1}$$

where $x(k) \in \mathbb{R}^n$, $u(k) \in \mathbb{R}$, $y(k) \in \mathbb{R}$ are respectively the state, the input, and the output of the system. Φ, Γ, C_1 are matrices of appropriate dimensions, h is an integer which denotes the amount of delay and $\Theta(k)$ denotes the initial condition. $\Theta(k)$ is generally available because it refers to past inputs which were applied to the system in past. D is a column matrix and $\rho(k) \in \mathbb{R}$. $D\rho(k)$ accounts for uncertainty which contains both matched and unmatched components.

Remark 4.2.1. *It is assumed that the system matrix Φ is non-singular. When Φ is singular then a possible way to circumvent this problem would be to slightly perturb the elements of Φ for ensuring non-singularity. Appropriate sampling time can also be chosen to ensure non-singularity of Φ when discrete-time model is obtained from its continuous time counterpart.*

It is also assumed that the pair (Φ, Γ) is controllable. The predictor can be used to convert the system (4.1) into a system which does not explicitly contains delay. A control input applied at the kth sampling instant becomes effective at the $(k+h)^{th}$ sampling instant due to the delay in the input. This situation demands that at the kth instant the controller should know the future value of the state at the $(k+h)^{th}$ instant. This can be accomplished by predicting the state from the plant dynamics.

Consider the deterministic part of (4.1) ($i.e$ $\rho(k) = 0$), then

$$\begin{aligned} x(k+2) &= \Phi x(k+1) + \Gamma u(k-h+1), \\ x(k+3) &= \Phi x(k+2) + \Gamma u(k-h+2), \\ x(k+3) &= \Phi^3 x(k) + \Phi^2 \Gamma u(k-h) + \\ &\quad \Phi \Gamma u(k-h+1) + \Gamma u(k-h+2), \\ &\vdots \\ &\vdots \\ x(k+h) = \hat{x}(k) &= \Phi^h x(k) + \sum_{i=-h+1}^{0} \Phi^{-i} \Gamma u(k+i-1). \end{aligned} \tag{4.2}$$

From (4.1) and (4.2), the system can be described in $\hat{x}$ coordinates as follows

$$\hat{x}(k+1) = \Phi\hat{x}(k) + \Gamma u(k) + \Phi^h D\rho(k). \tag{4.3}$$

It should be noted that the predicted state and the actual state are generated from the same dynamical system and there is always an 'h' time step difference between them. The value of the predicted state $\hat{x}$ at any sampling instant k, is the value of the actual state at the $(k+h)^{th}$ sampling instant. We can write the output in terms of the predicted state as follows

$$\hat{y}(k) = C_1\hat{x}(k). \tag{4.4}$$

4.3 Nonlinear Sliding Surface

This section discusses the design of a sliding surface for a class of uncertain SISO systems with a delay in the input. Without loss of generality the plant described by (4.3) and (4.4) can be transformed into the regular form by some transformation (because Γ is full rank) $z(k) = T_r\hat{x}(k)$ as

$$z_1(k+1) = \Phi_{11}z_1(k) + \Phi_{12}z_2(k) + \rho_1(k), \tag{4.5a}$$
$$z_2(k+1) = \Phi_{21}z_1(k) + \Phi_{22}z_2(k) + \Gamma_2 u(k) + \rho_2(k), \tag{4.5b}$$
$$\hat{y}(k) = Cz(k), \tag{4.5c}$$

where $z_1(k) \in \mathbb{R}^{n-1}$, $z_2(k) \in \mathbb{R}$, $C = C_1(T_r)^{-1}$, and Γ_2 is a scalar. $\rho_1(k) \in \mathbb{R}^{n-1}$ and $\rho_2(k) \in \mathbb{R}$ are unmatched and matched components of uncertainty $D\rho(k)$ respectively and

$$\begin{bmatrix} \rho_1(k) \\ \rho_2(k) \end{bmatrix} = T_r\Phi^h D\rho(k). \tag{4.6}$$

Define $z(k) := \begin{bmatrix} z_1(k) \\ z_2(k) \end{bmatrix}$. Similarly, in z coordinates let a desired trajectory be $z_d(k) := \begin{bmatrix} z_{1d}(k) \\ z_{2d}(k) \end{bmatrix}$.

Before we go for the expression of sliding surface, let us define:

$$c_1(k) := F - \Psi(\hat{y}(k))\Phi_{12}^T P(\Phi_{11} - \Phi_{12}F). \tag{4.7}$$

$$\begin{aligned} c^T(k) &:= \begin{bmatrix} c_1(k) & 1 \end{bmatrix} \\ &= \begin{bmatrix} F - \Psi(\hat{y}(k))\Phi_{12}^T P(\Phi_{11} - \Phi_{12}F) & 1 \end{bmatrix}. \end{aligned}$$

$$e_1(k) := z_1(k) - z_{1d}(k). \tag{4.8a}$$
$$e_2(k) := z_2(k) - z_{2d}(k). \tag{4.8b}$$

The sliding surface for the system in regular form is proposed as

$$s(k) := c^T(k)e(k), \tag{4.9a}$$

$$:= \begin{bmatrix} c_1(k) & 1 \end{bmatrix} \begin{bmatrix} z_1(k) - z_{1d}(k) \\ z_2(k) - z_{2d}(k) \end{bmatrix}, \tag{4.9b}$$

$$:= \begin{bmatrix} F - \Psi(\hat{y}(k))\Phi_{12}^T P(\Phi_{11} - \Phi_{12}F) & 1 \end{bmatrix} \begin{bmatrix} e_1(k) \\ e_2(k) \end{bmatrix}, \tag{4.9c}$$

In the above equation, F is chosen such that $(\Phi_{11} - \Phi_{12}F)$ has stable eigenvalues and dominant poles have a low damping ratio. $\Psi(\hat{y}(k))$ is a non-positive function used to change the damping ratio. P is a positive definite matrix, obtained from the solution of the following Lyapunov equation

$$P = (\Phi_{11} - \Phi_{12}F)^T P(\Phi_{11} - \Phi_{12}F) + W, \tag{4.10}$$

for some positive definite matrix W. Such a P exists because $(\Phi_{11} - \Phi_{12}F)$ is a stable matrix.

4.3.1 Selection of Nonlinear Function $\Psi(\hat{y}(k))$

The nonlinear term is used to change the system's closed loop damping ratio as the output approaches setpoint. The nonlinear function is chosen such that it changes from 0 to -β in a nonlinear manner. One possible choice of $\Psi(\hat{y}(k))$ is as follows

$$\Psi(\hat{y}(k)) = -\beta \frac{\left| |\hat{y}(k-1) - r(k)|^\alpha - |\hat{y}(0) - r(0)|^\alpha \right|}{|\hat{y}(0) - r(0)|^\alpha},$$

where $r(k)$ is the reference trajectory, $\beta > 0$ is used as a tuning parameter and α determines the rate of change of the damping ratio. The parameter β contributes to the final damping ratio of the system. In the above equation at $k = 0$, the predicted output $\hat{y}(k-1)$ can be approximated with $\hat{y}(0)$. It should be noted that the choice of $\Psi(\hat{y}(k))$ is not unique and any function with the above mentioned property can be used. Another possible choice is as follows

$$\Psi(\hat{y}(k)) = -\beta e^{-\bar{k}|\hat{y}(k-1) - r(k)|}, \tag{4.11}$$

where $\bar{k}$ is a positive constant. It easy to observe that as $\hat{y}(k)$ approaches reference $r(k)$, the function $\Psi(\hat{y}(k))$ approaches $-\beta$. The nonlinear function $\Psi(\hat{y}(k))$ has to satisfy the following condition

$$2\Psi(\hat{y}(k)) + \Psi(\hat{y}(k))\Phi_{12}^T P \Phi_{12} \Psi(\hat{y}(k)) \leq 0. \tag{4.12}$$

The following Lemma gives a necessary and sufficient condition of $\Psi(\hat{y}(k))$ so that condition (4.12) is satisfied.

Lemma 4.3.1. *Condition (4.12) is satisfied iff* $\Psi(\hat{y}(k)) \in [\frac{-2}{\gamma}, 0]$. *Where scalar* γ *is defined as* $\gamma := \Phi_{12}^T P \Phi_{12}$, *by definition, the matrix P is positive definite therefore* $\gamma > 0$.

Proof. Consider the condition

$$2\Psi(\hat{y}(k)) + \Psi(\hat{y}(k))\Phi_{12}^T P \Phi_{12} \Psi(\hat{y}(k)) \leq 0$$

Using the definition of γ

$$\begin{aligned}
&\Leftrightarrow 2\Psi(\hat{y}(k)) + \Psi^2(\hat{y}(k))\gamma \leq 0 \\
&\Leftrightarrow \Psi(\hat{y}(k))(2 + \Psi(\hat{y}(k))\gamma) \leq 0 \\
&\Leftrightarrow (2 + \Psi(\hat{y}(k))\gamma) \geq 0 \quad (\because \text{by definition } \Psi(\hat{y}(k)) \leq 0) \\
&\Leftrightarrow \Psi(\hat{y}(k)) \geq \frac{-2}{\gamma} \\
&\Leftrightarrow \Psi(\hat{y}(k)) \in [\frac{-2}{\gamma}, 0] \\
&(\because \gamma > 0 \text{ and by the definition } \Psi(\hat{y}(k)) \leq 0\ )
\end{aligned}$$

Remark 4.3.1. *As function* $\Psi(\hat{y}(k))$ *changes from* 0 *to* $-\beta$, *the above lemma also decides the range of parameter* β. *It is straight forward to verify that* $\beta \in [\frac{2}{\gamma}, 0]$ *ensures* $\Psi(\hat{y}(k)) \in [\frac{-2}{\gamma}, 0]$ *and using the above lemma, the condition (4.12) is satisfied.*

4.4 Control Law

The sliding surface discussed in the previous section is nonlinear and the surface parameters are changing at every sampling instant. The control law should be chosen in such a way that from any initial condition the system trajectory is attracted towards the sliding surface.

Consider the sliding surface

$$\begin{aligned}
s(k) &= c^T(k)e(k), \\
&= c^T(k)(z(k) - z_d(k)), \\
\Rightarrow s(k+1) &= c^T(k+1)\{z(k+1) - z_d(k+1)\} \\
&= c_1(k+1)\Phi_{11}z_1(k) + c_1(k+1)\Phi_{12}z_2(k) + c_1(k+1)\rho_1(k) + \\
&\quad \Phi_{21}z_1(k) + \Phi_{22}z_2(k) + \rho_2(k) + \Gamma_2 u(k) - c^T(k+1)z_d(k+1).
\end{aligned}$$

To reach the sliding surface in one sampling period, the reaching law as proposed in [98]

$$s(k+1) = 0,$$

which yields the control law:

$$u(k) = -(\Gamma_2)^{-1}\{c_1(k+1)\Phi_{11}z_1(k) + c_1(k+1)\Phi_{12}z_2(k) + c_1(k+1)\rho_1(k) + \Phi_{21}z_1(k) + \Phi_{22}z_2(k) + \rho_2(k) - c^T(k+1)z_d(k+1)\}. \tag{4.13}$$

This control law contains uncertain terms and therefore it cannot be implemented. To solve this problem, uncertain terms are replaced by its just previous instant values as proposed in [89]. By replacing $\rho_1(k)$ and $\rho_2(k)$ with $\rho_1(k-1)$ and $\rho_2(k-1)$ respectively, the control law (4.13) becomes

$$u(k) = -(\Gamma_2)^{-1}\{c_1(k+1)\Phi_{11}z_1(k) + c_1(k+1)\Phi_{12}z_2(k) + c_1(k+1)\rho_1(k-1) + \Phi_{21}z_1(k) + \Phi_{22}z_2(k) + \rho_2(k-1) - c^T(k+1)z_d(k+1)\}. \tag{4.14}$$

In the above control law, $\rho_1(k-1)$ and $\rho_2(k-1)$ can be computed from (4.5)

$$\rho_1(k-1) = z_1(k) - \Phi_{11}z_1(k-1) - \Phi_{12}z_2(k-1), \tag{4.15}$$

$$\rho_2(k-1) = z_2(k) - \Phi_{21}z_1(k-1) - \Phi_{22}z_2(k-1) - \Gamma_2 u(k-1). \tag{4.16}$$

Remark 4.4.1. *It is trivial to verify that the control law (4.14) ensures the existence of discrete-time sliding mode. The ultimate bounds of switching function and error are discussed in the next section. It should be noted that existence condition i.e,* $|s(k+1)| < |s(k)|$ *need to be satisfied when* $s(k)$ *is outside its ultimate bound.*

4.5 Stability of Motion

It is a well established fact that for an uncertain discrete-time system, ideal sliding motion is not possible (see [99, 12]). In this section we prove bounds of switching function $s(k)$ and state trajectory which result due to matched and unmatched perturbations.

4.5.1 Boundedness of $s(k)$

By the application of control law (4.14), the dynamics of switching function can be written as

$$s(k+1) = c_1(k+1)(\rho_1(k) - \rho_1(k-1)) + \rho_2(k) - \rho_2(k-1). \tag{4.17}$$

For bounded rate of change of disturbance, there exist positive scalars d_1 and d_2 such that following inequalities are satisfied.

$$c_{1m}\|\rho_1(k) - \rho_1(k-1)\| \le d_1, \tag{4.18a}$$

$$|rho_2(k) - \rho_2(k-1)| \le d_2, \tag{4.18b}$$

where c_{1m} is a positive scalar which satisfy $\|c_1(k+1)\| \le c_{1m}$. It implies

$$|s(k)| \le d_1 + d_2. \tag{4.19}$$

From the above equation it is clear that $s(k)$ is bounded by the maximum rate of change of disturbances. It should be noted that for constant disturbances, $s(k)$ remains precisely zero. By considering the availability of high speed digital signal processors and high end 32-bit microcontrollers, sampling rate can be chosen sufficiently small and this leads to a small boundary layer around $s(k)$. What follows next boundedness of $e_1(k)$ is discussed.

4.5.2 Boundedness of $e_1(k)$

From (4.9) and (4.8) it follows

$$e_2(k) = -c_1(k)e_1(k) + s(k). \tag{4.20}$$

From (4.5a) and (4.20) we can write the closed loop dynamics for the first subsystem as

$$\begin{aligned} e_1(k+1) = {} & \Phi_{11eq}e_1(k) + \Phi_{12}\Psi(\hat{y}(k))\Phi_{12}^T P\Phi_{11eq}e_1(k) + \Phi_{12}s(k) + \rho_1(k) + \\ & \Phi_{11}z_{1d}(k) + \Phi_{12}z_{2d}(k) - z_{1d}(k+1) \end{aligned} \tag{4.21}$$

where $\Phi_{11eq} = (\Phi_{11} - \Phi_{12}F)$. The pair $(\Phi,\ \Gamma)$ is controllable therefore there exist a control u_d such that [78, 40]

$$z_{1d}(k+1) = \Phi_{11}z_{1d}(k) + \Phi_{12}z_{2d}(k). \tag{4.22}$$

To simplify the notation, let us define

$$g(k) = \Phi_{12}s(k) + \rho_1(k). \tag{4.23}$$

Substituting (4.22) and (4.23) into (4.21)

$$e_1(k+1) = \Phi_{11eq}e_1(k) + \Phi_{12}\Psi(\hat{y}(k))\Phi^T P\Phi_{11eq}e_1(k) + g(k). \tag{4.24}$$

We prove the following lemmas which will subsequently be used for the proof of stability.

Lemma 4.5.1. *For vector $g(k)$ defined in (4.23)*

$$\|g(k)\| \le g_m.$$

Where

$$g_m := (d_1 + d_2)\|\Phi_{12}\| + \rho_{1m}\ , \tag{4.25}$$

in the above expression, $\rho_(1m)$ is a positive scalar such that $\|\rho_1(k)\| \le \rho_{1m}$, d_1 and d_2 are defined in (4.18).

Proof

$$g(k) = \Phi_{12}s(k) + \rho_1(k). \tag{4.26}$$

$$\Rightarrow \|g(k)\| \le \|\Phi_{12}s(k)\| + \|\rho_1(k)\| \tag{4.27}$$

As uncertainty is bounded, therefore there exists a scalar $\rho_{1m} > 0$ such that $\|\rho_1(k)\| \leq \rho_{1m}$.

Using the above bound of uncertainty and (4.19) it follows

$$\|g(k)\| \leq (d_1 + d_2)\|\Phi_{12}\| + \rho_{1m}.$$

So, $\|g(k)\| \leq g_m$ is proved.

Lemma 4.5.2. *For a vector* $\delta(k) := g^T(k)Pg(k)$,

$$\|\delta(k)\| \leq \delta_m,$$

where $\delta_m := \sigma_{max}(P)g_m^2$. $\sigma_{max}(P)$ *represents maximum eigen value of* P*; since* $P > 0$ *it implies* $\sigma_{max}(P) > 0$.

Proof. Using the Rayleigh principle

$$\begin{aligned} &\|\delta(k)\| \leq \sigma_{max}(P)\|g(k)\|^2. \\ \text{Using bound of } g(k) \text{ from Lemma 4.5.1} \quad &\|\delta(k)\| \leq \sigma_{max}(P)g_m^2. \\ \text{Therefore} \quad &\|\delta(k)\| \leq \delta_m. \end{aligned}$$

Lemma 4.5.3. $\|\Psi(\hat{y}(k))\| \leq \beta$

Proof. Proof is obvious from the definition of $\Psi(\hat{y}(k))$.

Lemma 4.5.4. *A vector* $\alpha(k) := g^T(k)P\Phi_{11eq} + g^T(k)P\Phi_{12}\Psi(\hat{y}(k))\Phi_{12}^T P\Phi_{11eq}$ *is bounded by*

$$\|\alpha\| \leq \alpha_m. \tag{4.28}$$

Where $\alpha_m = g_m\|P\|\|\Phi_{11eq}\|(1 + \beta\|P\|\|\Phi_{12}\|^2)$, *the matrix* P *is obtained from the solution of (4.10). All other matrices are discussed in the previous sections.*

Proof

$$\begin{aligned} \alpha &= g^T(k)P\Phi_{11eq} + g^T(k)P\Phi_{12}\Psi(\hat{y}(k))\Phi_{12}^T P\Phi_{11eq}, \\ \Rightarrow \|\alpha\| &\leq \|g^T(k)P\Phi_{11eq}\| + \|g^T(k)P\Phi_{12}\Psi(\hat{y}(k))\Phi_{12}^T P\Phi_{11eq}\|, \\ &\leq \|g(k)\|\|P\|\|\Phi_{11eq}\| + \beta\|g(k)\|\|P\|^2\|\Phi_{12}\|^2\|\Phi_{11eq}\|, \\ &= \|g(k)\|\|P\|\|\Phi_{11eq}\|(1 + \beta\|P\|\Phi_{12}\|^2) \\ \text{From Lemma 4.5.1,} \quad \|\alpha\| &\leq g_m\|P\|\|\Phi_{11eq}\|(1 + \beta\|P\|\Phi_{12}\|^2), \\ \Rightarrow \|\alpha\| &\leq \alpha_m. \end{aligned}$$

The third inequality follows from the use of Lemma 4.5.3 and standard arguments.

The boundedness of $e_1(k)$ resulting from (4.24) is proved in the following theorem.

Theorem 4.5.1. *If* $(\Phi_{11} - \Phi_{12}F)$ *is stable and* $\Psi(\hat{y}(k))$ *satisfies (4.12) then the trajectory* $e_1(k)$ *of subsystem (4.24) uniformly ultimately bounded with radius* $r = \varepsilon + |\frac{\alpha_m - \sqrt{\alpha_m^2 + \sigma_{min}(W)\delta_m}}{\sigma_{min}(W)}|$.

Where α_m is defined in Lemma 4.5.4, δ_m is defined in Lemma 4.5.2, $\sigma_{min}(W)$ is the smallest eigenvalue of matrix W. The matrix $P > 0$ is obtained from the solution of (4.10) for some $W > 0$.

Proof. Let a Lyapunov function for system (4.24) be defined as

$$V(k) = e_1^T(k)Pe_1(k).$$

An increment of $V(k)$ becomes

$$\Delta V(k) = V(k+1) - V(k).$$

It implies

$$\begin{aligned}
\Delta V(k) &= e_1^T(k+1)Pe_1(k+1) - e_1^T(k)Pe_1(k), \\
&= \{\Phi_{11eq}e_1(k) + \Phi_{12}\Psi(\hat{y}(k))\Phi_{12}^T P\Phi_{11eq}e_1(k) + g(k)\}^T \\
&\quad \times P\{\Phi_{12}\Psi(\hat{y}(k))\Phi_{12}^T P\Phi_{11eq}e_1(k) + \Phi_{11eq}e_1(k) + g(k)\} - e_1(k)^T Pe_1(k), \\
&= e_1^T(k)\Phi_{11eq}^T P\Phi_{12}\Psi(\hat{y}(k))\Phi_{12}^T P\Phi_{11eq}e_1(k) + e_1^T(k)\Phi_{11eq}^T e_1(k) - e_1^T(k)Pe_1(k) + \\
&\quad e_1^T(k)\Phi_{11eq}^T P\Phi_{12}\Psi(\hat{y}(k))\Phi_{12}^T P\Phi_{11eq}e_1(k) + \\
&\quad e_1^T(k)\Phi_{11eq}^T P\Phi_{12}\Phi_{12}^T\Phi_{12}\Psi(\hat{y}(k))\Phi_{12}^T P\Phi_{11eq}e_1 + 2g^T(k)P\Phi_{11eq}e_1(k) + \\
&\quad 2g^T(k)P\Phi_{12}\Psi(\hat{y}(k))\Phi_{12}^T P\Phi_{11eq}e_1(k) + g^T(k)Pg(k), \\
&= -e_1^T(k)We_1(k) + \\
&\quad e_1^T(k)\Phi_{11eq}^T P\Phi_{12}\{2\Psi(\hat{y}(k)) + \Psi(\hat{y}(k))\Phi_{12}^T P\Phi_{12}\Psi(\hat{y}(k))\}\Phi_{12}^T P\Phi_{11eq}e_1 + \\
&\quad 2g^T(k)P\Phi_{11eq}e_1(k) + 2g^T(k)P\Phi_{12}\Psi(\hat{y}(k))\Phi_{12}^T P\Phi_{11eq}e_1(k) + g^T(k)Pg(k).
\end{aligned}$$

To simplify the notations, let $e_1^T(k)\Phi_{11eq}^T P\Phi_{12} = M^T$, and using the definition of α and δ from Lemma 4.5.4 and Lemma 4.5.2 respectively

$$\begin{aligned}
\Delta V(k) = M^T\{2\Psi(\hat{y}(k)) + \Psi(\hat{y}(k))\Phi_{12}^T P\Phi_{12}\Psi(\hat{y}(k))\}M - e_1^T(k)We_1(k) + \\
2\alpha e_1(k) + \delta(k).
\end{aligned}$$

From the condition (4.12) we can now write

$$\Delta V(k) \leq -e_1^T(k)We_1(k) + 2\alpha e_1(k) + \delta(k).$$

From Lemma 4.5.4 and using standard arguments it follows

$$\Delta V(k) \leq -\sigma_{min}(W)\|e_1\|^2 + 2\|e_1\|\alpha_m + \|\delta(k)\|.$$

Using the bound of $\delta(k)$ from Lemma 4.5.2

$$\Delta V(k) \leq -\sigma_{min}(W)\|e_1\|^2 + 2\|e_1\|\alpha_m + \delta_m.$$

It can be easily seen that if

$$\|e_1\| \geq \varepsilon + |\frac{\alpha_m - \sqrt{\alpha_m^2 + \sigma_{min}(W)\delta_m}}{\sigma_{min}(W)}|, \tag{4.29}$$

then $\Delta V(k) < 0$, where ε is an arbitrarily small constant. This implies that trajectory will enter in a ball which has center at origin and has radius

$$r = \varepsilon + |\frac{\alpha_m - \sqrt{\alpha_m^2 + \sigma_{min}(W)\delta_m}}{\sigma_{min}(W)}|. \tag{4.30}$$

So the nonlinear sliding surface (4.9) leads to the stable closed loop dynamics, thus the nonlinear surface is stable and the Theorem is proved.

Remark 4.5.1. *It has been proved that $s(k)$ and $e_1(k)$ both are bounded, from this it is straightforward to deduce boundedness of $e_2(k)$ by using (4.20).*

Remark 4.5.2. *During the sliding mode, the dynamics of the system are decided by the closed loop subsystems described by (4.17) and (4.24). The subsystem in (4.24) is stable ($e_1(k)$ remains bounded) for any non-positive value of $\Psi(\hat{y}(k))$ which respects (4.12). The function $\Psi(\hat{y}(k))$ changes from 0 to $-\beta$ as the output tracks the reference signal. For any intermediate value of Ψ, the closed loop system (4.24) is stable. As it is proved in [70] that the introduction of this function changes the damping ratio of the system from an initial value ζ_1 to a final value ζ_2, where $\zeta_2 > \zeta_1$. Initially, when $\Psi(\hat{y}(k)) = 0$, the damping ratio is contributed by F which is designed for a low damping ratio. When the output gets closer to the reference, $\Psi(\hat{y}(k))$ contributes significantly to the damping ratio of the system. At steady state, the value of $\Psi(\hat{y}(k))$ becomes $\Psi(\hat{y}(k)) \approx -\beta$ and then the subsystem (4.24) can be written as*

$$e_1(k+1) = (\Phi_{11} - \Phi_{12}F - \beta\Phi_{12}\Phi_{12}^T P(\Phi_{11} - \Phi_{12}F))e_1(k) + g(k). \tag{4.31}$$

The above equation determines the final damping ratio, and therefore the parameter β and matrix P can be designed such that the dominant poles of (4.31) have a desired damping ratio. (4.31) can be written as

$$e_1(k+1) = (\Phi_{11} - (F + \beta\Phi_{12}^T P(\Phi_{11} - \Phi_{12}F))\Phi_{12})e_1(k) + g(k).$$

Let a desired final damping ratio be ζ_2 and a required gain be K_2, which can be computed by pole placement technique. Thus

$$K_2 = F + \beta\Phi_{12}^T P(\Phi_{11} - \Phi_{12}F), \tag{4.32}$$

and this simplifies to

$$\Phi_{12}^T P = \frac{(K_2 - F)(\Phi_{11} - \Phi_{12}F)^{-1}}{\beta}. \tag{4.33}$$

The above equation can be solved for $P = P^T > 0$. There are many solutions because $0.5n(n-1)$ elements are available for selection. After finding the matrix P, the matrix W can be calculated using (4.10). If W is not at least positive semi definite then choose another $P > 0$ and repeat the procedure. Parameter β can also be altered to get the desired final damping ratio. However, this procedure may require several iterations. To automate the tuning procedure, the following section discusses an LMI based tuning algorithm.

4.5.3 LMI Based Tuning Algorithm

In the above remark an iterative procedure is mentioned to tune parameters of non-linear surface. To obtain desired final damping ratio, (4.33) should satisfy and to ensure stability (4.10) must be satisfied. The objective is to search a matrix P such that (4.10) is satisfied and (4.33) satisfies as closely as possible. These constraints can be posed as a linear matrix inequalities and can be easily solved by many commercially available packages. Let us define

$$H_1 := \Phi_{12}^T P - \frac{(K_2 - F)(\Phi_{11} - \Phi_{12}F)^{-1}}{\beta}. \tag{4.34}$$

The above equality can be converted into an inequality so that the above equation can satisfy as closely as possible. Our objective is to search a matrix P to minimize some $\mu > 0$ such that

$$\| H_1 \| \leq \mu. \tag{4.35}$$

The above nonlinear inequality can be converted into a linear inequality by using Schur's lemma as follows

$$\begin{bmatrix} \mu\, I & H_1 \\ H_1^T & \mu\, I \end{bmatrix} > 0. \tag{4.36}$$

Therefore we can cast an optimization problem as follows: Minimize μ such that

$$P > 0, \tag{4.37a}$$

$$(\Phi_{11} - \Phi_{12}F)^T P(\Phi_{11} - \Phi_{12}F) - P < 0\,, \tag{4.37b}$$

$$\begin{bmatrix} \mu\, I & H \\ H^T & \mu\, I \end{bmatrix} > 0. \tag{4.37c}$$

The above LMI can be easily solved for P by minimizing ε. It should be noted that the freedom to choose β can also be used to minimize the objective function. This approach automate the tuning procedure and reduces the number of manual iteration required for tuning.

Remark 4.5.3. *In Theorem 4.5.1, we have proved stability of the system in z-coordinates which ensures stability of $\hat{x}(k)$ because $z(k)$ and $\hat{x}(k)$ are related through a linear transformation. Because of the input-delay, the actual state vector $x(k)$ lags behind the predicted state vector $\hat{x}(k)$ by an h-sampling instants. $x(k)$ will change* *exactly the same way as $\hat{x}(k)$ changes with a constant time lag of h sampling periods. With this relation between $x(k)$ and $\hat{x}(k)$ we can conclude that the stability of $\hat{x}(k)$ ensures the stability of $x(k)$.*

4.6 Illustrative Example

In this section, a numerical example is presented to illustrate the proposed method. The following qualitative control objectives are to be achieved:

- System should have small overshoot.
- System should track step trajectory with low settling time
- System should be robust.

Simulation results are presented for different fixed-damping ratio sliding surfaces and compared with the proposed nonlinear sliding surface. Two different cases are considered in first case we simulate without disturbance to show improvement in performance over linear surfaces (linear in the predicted state) and in the second case disturbance is added to show robustness property of the proposed nonlinear surface.

Consider the following system matrices

$$\Phi = \begin{bmatrix} 1.0000 & 0.0900 & 0.0069 & 0.0002 \\ 0 & 1.0000 & 0.1336 & 0.0073 \\ 0 & 0 & 0.3851 & 0.0652 \\ 0 & 0 & -11.6909 & 0.2863 \end{bmatrix}$$

$$\Gamma = \begin{bmatrix} 0.0007 \\ 0.0162 \\ -0.0057 \\ -0.1082 \end{bmatrix} \qquad C_1 = \begin{bmatrix} 1 \; 0 \; 0 \; 0 \end{bmatrix}$$

amount of time delay $h = 5$.

Sampling time is $\tau = 0.09 sec$.

CASE-I System without disturbance

To show improvement in performance, first deterministic system is considered (i.e $\rho(k) = 0$). In the following nonlinear surface design and corresponding control law is presented

Step 1: Transform the system into a delay free system by use of the predictor

$$\hat{x}(k) = \Phi^5 x(k) + \sum_{i=-4}^{0} \Phi^{-i} \Gamma u(k+i-1). \tag{4.38}$$

to give

$$\hat{x}(k+1) = \Phi \hat{x}(k) + \Gamma u(k)$$

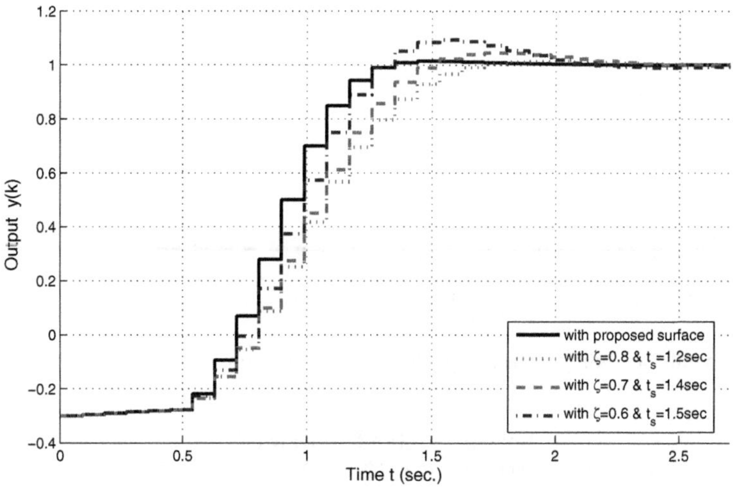

Fig. 4.1 Response of output y(k) with different sliding surfaces

Step 2: Transform the system into regular form by the appropriate T_r matrix. With $z = T_r\hat{x}$ system matrices in regular form becomes

$$\Phi_{11} = \begin{bmatrix} 0.9609 & -1.5659 & 0.0753 \\ 0.0192 & 0.9787 & 0.0047 \\ 0.0811 & -0.3050 & 1.0278 \end{bmatrix}$$

$$\Phi_{12} = \begin{bmatrix} -0.1748 \\ 0.0987 \\ -0.0532 \end{bmatrix}$$

$\Phi_{21} = \begin{bmatrix} -0.1914 & -11.5305 & 0.5689 \end{bmatrix}$
$\Phi_{22} = -0.2961 \quad \Gamma_2 = -0.1096.$

The required trajectory in $\hat{x}$ coordinates is $\hat{x}_d = [1 \ \ 0 \ \ 0 \ \ 0]^T$ and $r(k) = 1$. By using the transformation matrix T_r required trajectory and output can be found in $z-$cordinates.

Step 3: Design the nonlinear sliding surface. The nonlinear sliding surface is composed of a constant and a nonlinear term. Initially the nonlinear term is zero, and therefore the constant term decides the initial damping ratio (ζ_1) and settling time. For an initial settling time of $t_{s1} = 9sec.$ and an initial damping ratio of $\zeta_1 = 0.1$, matrix F can be found as $F = \begin{bmatrix} 0.2396 & 5.1593 & -6.7888 \end{bmatrix}$.

By solving LMIs in (4.37) for final settling time $t_{s2} = 3.311 \ \ Sec.$ and final damping ratio $\zeta_1 = 0.27$, the matrix P is obtained as

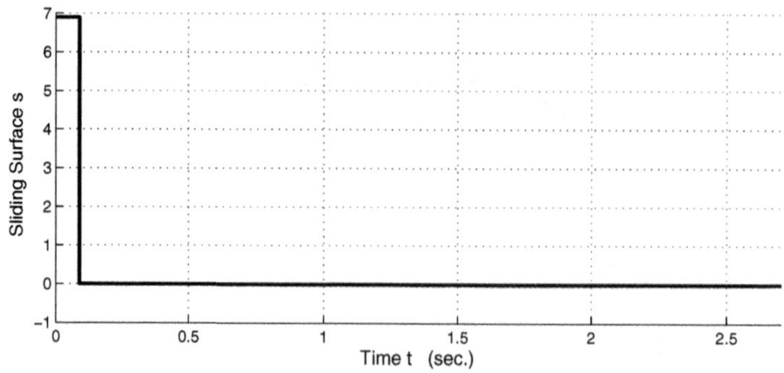

Fig. 4.2 Plot of nonlinear sliding surface in predicted state $\hat{x}$ with time.

$$P = 10^3 \begin{bmatrix} 0.0825 & -0.0696 & -0.1136 \\ -0.0696 & 0.1012 & 0.2579 \\ -0.1136 & 0.2579 & 1.1930 \end{bmatrix}.$$

Matrix P is positive definite as required.

Function $\Psi(\hat{y}(k))$, with $\beta = 0.1254$ and $\alpha = 3$ is given as follows

$$\Psi(\hat{y}(k)) = -0.1254 \frac{|\,|\hat{y}(k-1) - r(k)|^3 - |\hat{y}(0) - r(0)|^3|}{|\hat{y}(0) - r(0)|^3}.$$

From the above values, the switching function $s(k)$ can be computed.

Table 4.1 Settling time of output y(k) obtained by the proposed sliding surface versus surfaces with diff. ζ

Type of sliding surface	Peak Overshoot (%)	Settling Time t_s(sec.)
surface with ζ=0.6 and $t_s = 1.5$	9.4 %	1.98
surface with ζ=0.7 and $t_s = 1.4$	4.8%	2.16
surface with ζ=0.8 and $t_s = 1.2$	1.5%	1.62
Proposed surface	1.4%	1.26

Step 4: Design the control law.

The control law given in (4.14) can be implemented because all parameters are known.

4.6.1 Design of Linear Sliding Surface

The performance of the proposed nonlinear sliding surface is now compared with controllers designed with different linear sliding surfaces. The system output response is plotted for different sliding surfaces and the proposed sliding surface. The

parameters of the linear sliding surface(ζ and t_s) are chosen based on the values of these parameters taken at different instants when the nonlinear sliding surface is used. When the nonlinear sliding surface is used, poles of the closed loop system change as the output approaches the reference. As a consequence the damping ratio and settling time also change from their initial values ζ_1 and t_{s1} to final values ζ_2 and t_{s2}. The following three different linear sliding surfaces
Linear surface-1 with $\zeta = 0.6, t_s = 1.5$
Linear surface-2 with $\zeta = 0.7, t_s = 1.4$
Linear surface-3 with $\zeta = 0.8, t_s = 1.2$
are designed It should be noted that all these values of ζ and t_s occur at different instants when the nonlinear sliding surface is used. The sliding surface is given as follows

$$\begin{aligned} s_1 &= \bar{c}^T z \\ &= \begin{bmatrix} F & 1 \end{bmatrix} \begin{bmatrix} z_1 - z_{1d} \\ z_2 - z_{2d} \end{bmatrix}. \end{aligned} \tag{4.39}$$

F can be computed for different damping ratios and settling times. The control law for each linear sliding surface is given as follows

$$u(k) = -(\bar{c}^T T_r \Gamma)^{-1} \{\bar{c}^T T_r \Phi \hat{x}(k) - \bar{c}^T T_r x_d(k+1)\}$$

The above equation is derived based on equivalent control and therefore one step reaching is possible.

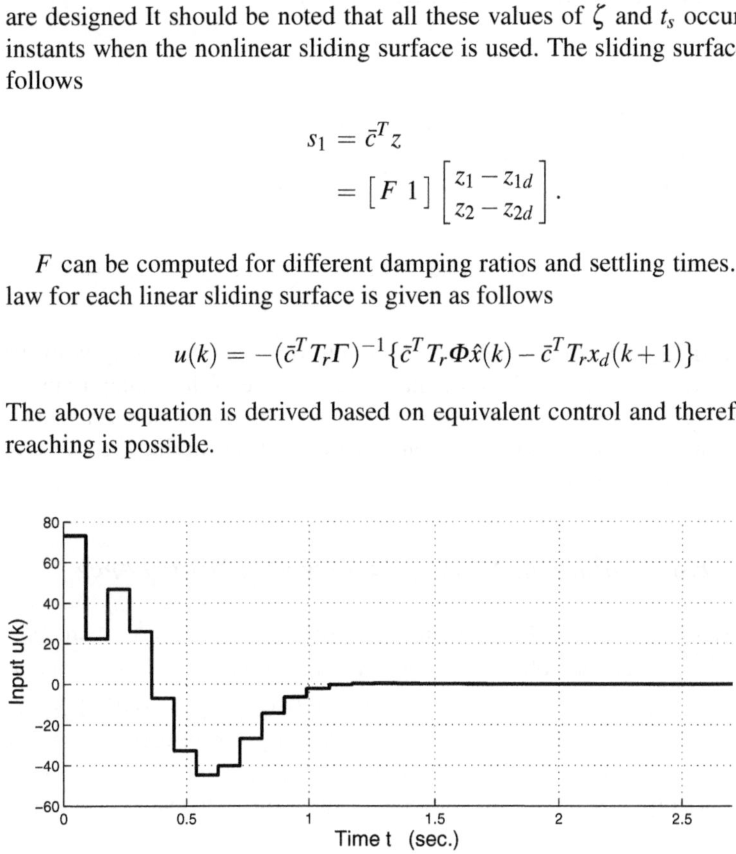

Fig. 4.3 Plot of input when nonlinear sliding surface is used

4.6.2 Comparison with Different Linear Sliding Surfaces

The response of the output $y = x_1$ is plotted for the nonlinear sliding surface and different linear sliding surfaces in Fig. 4.1. The plot clearly shows that with the nonlinear sliding surface the performance improves significantly. With the proposed surface the output settles in 1.26 seconds with negligible overshoot. The smallest improvement in settling time is 22%. Furthermore, the example illustrates very clearly

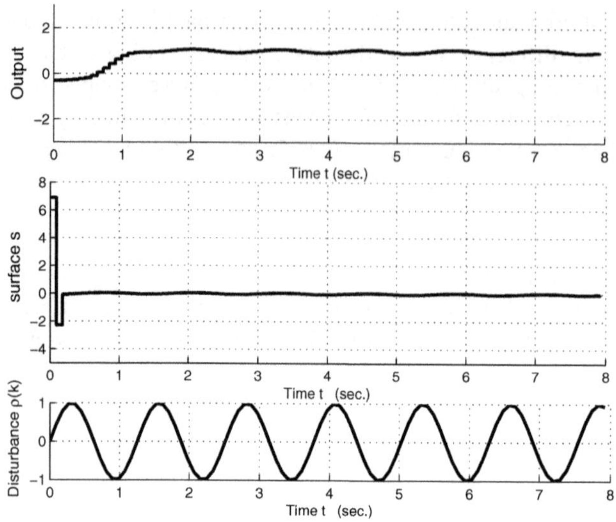

Fig. 4.4 Plot of output, surface and disturbance when nonlinear sliding surface is used

that peak overshoot and settling time can be minimized simultaneously with the nonlinear sliding surface. Fig. 4.2 shows that the amplitude of the sliding function in predicted state $\hat{x}$ becomes zero in one sampling period and remains zero thereafter with d(k)=0. Fig. 4.3 shows the input and confirms that chattering is eliminated because of equivalent control.

4.6.3 CASE-II Nonlinear Sliding Surface with Disturbance

Plant is perturbed by external unmatched disturbance $\rho = sin(5k)$ with $D = \left[0.03\ 0\ 0\ 0.03\right]^T$ to validate robustness property of the proposed surface. In z coordinates, using (4.6) $\rho_1 = sin(5k)\left[-0.0013\ 0.0001\ 0.0305\right]^T$ $\rho_2 = 0.0204sin(5k)$ Following bounds are computed using the formulation given in earlier Sections. $c_{1m} = 37.8413, d_1 = 5.7701, d_2 = 0.1020, g_m = 1.2499, \rho_{1m} = 0.0305, \delta_m = 194.24,$ $\alpha_m = 140.73$. Using these parameters, radius of ultimate boundedness computed from (4.30) as $r = 0.6884$.

Following the procedure described in Section-4.4, control law as per (4.14) is designed.

Fig. 4.4 shows plot of output, surface and disturbance when the plant is subjected to a disturbance. The closer look at the output plot reveled that there are some oscillations in output but of very small amplitude. From the same plot it can be verified that switching function remains in the vicinity of zero and thus the matched disturbances are rejected effectively and due to unmatched disturbance output and switching function remains bounded.

4.7 Conclusion

In this chapter, a nonlinear sliding surface has been proposed for improving the performance of an uncertain system with a time-delay in the input. The proposed control algorithm with the nonlinear surface is able to achieve low overshoot and low settling time simultaneously. Robustness and high performance both can be achieved with the proposed scheme. Disturbance rejection property of the proposed method is also shown. From Table-1 we can observe that with the illustrative example the minimum improvement in the settling time is 22% with negligible overshoot. Also, we note it is straight-forward to extend our results to the output feedback case.

Chapter 5
Integral Sliding Mode Based Composite Nonlinear Feedback Control

5.1 Introduction

In this chapter we discuss the problem of improving performance yet preserving the invariance towards the matched disturbances from a different perspective. In the previous chapters, we proposed various schemes in which the nonlinear surfaces are designed for different types of systems for the improvement of performance. In this chapter we propose a nonlinear surface which considers actuator saturation and the elimination of the reaching phase with improvement in the performance. For any practical system, actuator output can not take any amplitude. Actuator capacity is always bounded, therefore it is necessary to consider the effect of saturation actuator a *priori*. In conventional sliding mode, the motion of the trajectory is constrained to lie in an $(n-m)$ dimensional manifold with a discontinuous control action. Here m is the number of inputs and n is the order of the system. The motion of the trajectory from the initial condition towards sliding surface until it hits the sliding surface is called the reaching phase. During the reaching phase, the system is not robust and even matched disturbances can affect the system performance. To solve this problem, in [100], an integral sliding mode (ISM) concept is proposed. An integral term is incorporated in the sliding manifold, this guarantees that the system trajectories will start in the manifold right from the beginning thus, the reaching phase is eliminated; and the system becomes invariant towards matching perturbation right from the beginning. The main idea behind the ISM controller is to define the control law as a sum of a nominal control and a discontinuous control. Nominal control takes care of the nominal plant dynamics and the discontinuous control rejects the disturbances. The nominal control can be of any form which is able to follow the reference trajectory within a given accuracy. In this work we have taken Composite Nonlinear Feedback (CNF) controller, which is based on variable damping ratio, as a nominal controller along with the ISM controller to reject disturbance.

To achieve high performance and robustness, here we subtly combine CNF - to achieve high performance with saturated actuator- and the integral sliding mode to ensure invariance against disturbances.

B. Bandyopadhyay et al.: Sliding Mode Control Using Novel Sliding Surfaces, LNCIS 392, pp. 83–95.
springerlink.com

The main objective of this chapter is to achieve high performance in an uncertain environment which is a more realistic case. To verify the proposed theory, experimental investigation is carried out on a laboratory based servo position control system under the influence of a sustained disturbance. This chapter includes some results published in [6]. The rest of the chapter is organized as follows. In the next section we present the CNF controller for the plant with matched uncertainty. We have modified the linear part of CNF controller to avoid saturation with ISM controller. Section 3 presents an integral sliding mode with CNF controller and the proof of stability with the combined control law. In Section 4 a step by step approach is given to design the controller based on the proposed theory with experimental results for a servo position control problem followed by the concluding Section 5.

5.2 Composite Nonlinear Feedback Control

In this section the CNF controller is reviewed. It is shown that by restricting the maximum amplitude of a reference signal to be tracked, how ISMC can be combined with the CNF. Discussion in this section will go parallel to [21] with the modification that the maximum amplitude of the reference signal to be tracked is affected by the maximum value of the uncertainty. Consider the nominal plant with the uncertainty

$$\dot{x} = Ax + Bsat(u) + B\rho_d(x,t). \tag{5.1}$$

$$y = Cx. \tag{5.2}$$

Where $x \in \mathbb{R}^n$, $u \in \mathbb{R}$, $y \in \mathbb{R}$ are , respectively, the state, the input, the output of the system. A, B, C are the matrices of appropriate dimensions. $\rho(x,t)$ is bounded matched uncertainty and

$$|\rho(x,t)| \leq \rho_{dmax}, \tag{5.3}$$

where $\rho_{dmax} > 0$ is the maximum amplitude of disturbance.

Function sat:$R \to R$ represents actuator saturation which is defined as follows

$$sat(u) = sign(u) \times min u_{max}, |u|. \tag{5.4}$$

Where u_{max} is the maximum value of the input. Here single input case is considered to show the effectiveness of the concept in a simple manner. However, it should be noted that the concept can be extended for a multi input case also. In this work the following assumptions are made

1. Pair(A, B) is controllable.
2. (A, B, C) has no zeros at $s = 0$.
3. Maximum value of disturbance ρ_{dmax} is known .

Control Objective: Controlled output y should track the input command r as fast as possible in the presence of disturbances and uncertainty in the plant without experiencing a large overshoot.

As mentioned earlier that the CNF is made of two parts, linear and nonlinear. For system (5.1) linear feedback control law is given as follows

$$u_L = Fx + G_1 r. \tag{5.5}$$

F is chosen such that $(A+BF)$ is an asymptotically stable matrix, and it should have a small damping ratio. Furthermore the maximum amplitude of a reference signal which can be tracked is determined in Lemma 5.3.1. The matrix G_1 is defined as

$$G_1 = -[C(A+BF)^{-1}B]^{-1}. \tag{5.6}$$

This can be found by finding d.c gain from y to r. When y tracks r, the state x will take a new steady state value x_e which is given as

$$x_e = -(A+BF)^{-1}BG_1 r = G_e r, \tag{5.7}$$

where $G_e = -(A+BF)^{-1}BG_1$. This can be found by finding d.c gain between x and r. Let us make co-ordinate transformation as

$$\tilde{x} = x - x_e, \tag{5.8}$$

where $\tilde{x}$ is the error between the actual state and the desired state. The linear control law (5.5) in error coordinate becomes

$$u_L = F\tilde{x} + Hr, \tag{5.9}$$

where $H = [I - F(A+BF)^{-1}B]G_1$.

5.2.1 Nonlinear Control

The gain F is so chosen that $(A+BF)$ has a low damping ratio. To increase the damping ratio, a nonlinear control is added to the linear controller which is given as follows

$$u_N = \Psi(r,y)B^T P(x - x_e), \tag{5.10}$$

where $\Psi(r,y)$ is any non positive function locally lipschitz in y. The nonlinear control is used to change the system's closed loop damping ratio as the output approaches the desired reference command signal r. Initial choice for the selection of $\Psi(r,y)$ given in [70] is modified in [21] to get an extra degree of freedom. The nonlinear function should be chosen such that it changes from 0 to -β where $\beta > 0$. One possible choice of function as suggested in [21] as

$$\Psi(r,y) = -\frac{\beta}{1-e^{-1}}\left(e^{-|1-(y-y_0)/(r-y_0)|} - e^{-1}\right), \tag{5.11}$$

where $y_0 = y(0)$, and β is used as a tuning parameter.

From (5.5) and (5.10) the CNF control law can be given by combining the linear and nonlinear components as

$$u_0 = u_L + u_N = Fx + G_1 r + \Psi(r,y)B^T P(x - x_e). \tag{5.12}$$

5.3 Integral Sliding Mode Control

In this section the design of another nonlinear component which is based on integral sliding mode is presented. The objective is to cancel the model uncertainty and disturbances by combining CNF with ISM. The control component from ISM is added to the CNF control law (5.12) to ensure invariance against disturbances and the stability with this combined control law is proved.
Integral sliding manifold proposed in [100, 19] as

$$s(x,t) = G\{x(t) - x_0 - \int_0^t (Ax(\tau) + Bu_0(\tau))\,d\tau\,\}. \tag{5.13}$$

In the above sliding surface equation u_0 can be of any form. The u_0 when substituted in nominal plant(i.e (5.1) without disturbance term $\rho(x,t)$) gives nominal trajectory. In the above equation the term $\int_0^t (Ax(\tau) + Bu_0(\tau))\,d\tau$ represents desired trajectory and $x(t)$ is actual trajectory. So sliding surface can be seen as the difference between actual and the desired trajectory projected on G. The difference occurs due to disturbances. By using an appropriate control action this difference can be nullified and actual trajectory tracks the required one. The objective of ISMC is to force the sliding motion on $s(x,t) = 0$, therefore it works as a disturbance observer as proposed in [100]. Let this discontinuous control be u_{N1}. Therefore the total control to be applied becomes

$$u = u_0 + u_{N1} = u_L + u_N + u_{N1}. \tag{5.14}$$

The above control law will be refereed as ISM-CNF control law in further discussion.

Here the discontinuous control u_{N1} added to the CNF control, is required for enforcing the sliding motion. Fig. 5.1 shows the block diagram of the proposed scheme. To find the discontinuous control law, construct Lyapunov function as

$$V(x,t) = 0.5s^T(x,t)s(x,t). \tag{5.15}$$

By following the standard procedure as proposed in [19, 100], the usual choice for discontinuous control is

$$u_{N1} = -M(x,t)sign((GB)^T s(x,t)). \tag{5.16}$$

Where $M(x,t) \geq \rho_{max}$ is a positive definite function and calculated based on the maximum disturbance bounds. With this discontinuous control law negative definiteness of Lyapunov function can be easily proved. The control law in (5.16) is discontinuous and the actual effect of such a controller on a given plant is equal to the average of the control action, the so-called equivalent control [99]. For the analysis purpose, u_{N1} will be replaced with its equivalent value

$$(u_{N1})_{eq} = u_{eq} = -\rho(x,t) \tag{5.17}$$

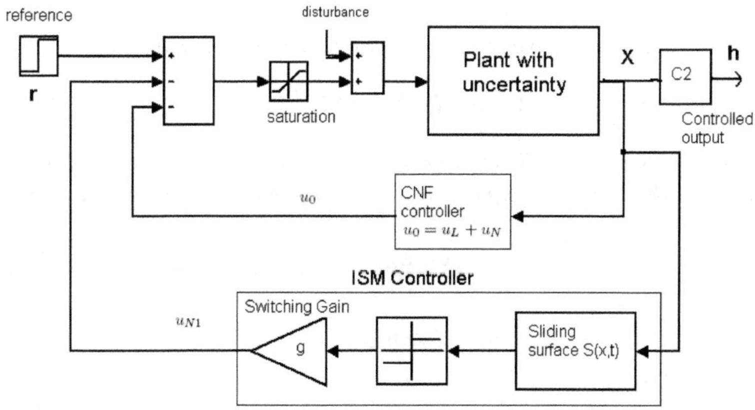

Fig. 5.1 Block diagram with combined ISM-CNF control law

as proposed in [16, 96, 99]. It can be observed that the equivalent control directly can not be implemented because it contains unknown disturbance term. For practical applications where to implement (5.16) is not suitable, output of a first order linear filter with discontinuous control as its input can be used as suggested in [96]. The low pass filter to obtain equivalent control is also proposed and successfully implemented in [105] in conventional SMC framework.

With limited control input it is necessary to determine the maximum amplitude of the reference signal which can be tracked and the maximum amplitude of disturbance which can be handled. The following lemma addresses these concerns. The objective of this lemma is to find the maximum amplitude of reference signal to be tracked and the maximum disturbance which can be handled. It will be proved in Theorem-5.3.1 that the amplitude of u_N does not affect the stability so long it satisfies properties discussed in section 5.2. Therefore the inputs u_L and u_{N1} are considered. Let us define

$$\bar{u} = u_L + u_{N1} \tag{5.18}$$

Remark 5.3.1. *It should be again emphasized that the purpose of this lemma is to determine the maximum amplitude of reference signal which can be tracked and the maximum amplitude of disturbance which can be tolerated. The stability with the combined control law (5.14) is proved in Theorem-5.3.1. The maximum amplitude of the reference signal which can be tracked and the maximum disturbance which can be tolerated are independent of u_N therefore in the following they are analysed for the control law (5.18).*

Lemma 5.3.1. *Let the real symmetric matrix $P > 0$ be the solution of the following Lyapunov equation*

$$(A+BF)^T P + P(A+BF) = -W, \tag{5.19}$$

for positive definite matrix W. Such a P exist, because $(A+BF)$ *is a stable matrix. Let* $c_\delta > 0$ *be the largest positive scalar satisfying the following condition*

$$| F\tilde{x} | \leq u_{max}(1-\delta)\forall \tilde{x} \in X_\delta = \{\tilde{x}|\tilde{x}^T P\tilde{x} \leq c_\delta\}, \tag{5.20}$$

where $\delta \in (0,1)$*. Then the feedback control law (5.18) makes the system output to track asymptotically a step command input of amplitude r, when initial error and r satisfy*

$$\forall \tilde{x} \in X_\delta, \tag{5.21}$$

$$|Hr| \leq \delta_1 u_{max}, \tag{5.22}$$

where, $0 < \delta_1 < \delta$ *and*

$$| (\delta - \delta_1)u_{max} | = \rho(x,t)_{max} \tag{5.23}$$

The above condition ensures that

$$u_L \leq \varepsilon u_{max}, \tag{5.24}$$

and

$$0 < \varepsilon = (1-\delta+\delta_1) < 1. \tag{5.25}$$

Remark 5.3.2. *In the statement of this lemma control efforts* $u_{max}\delta$ *have been redistributed on disturbance and Hr unlike [21]. The redistribution is done to ensure the stability of closed loop system with uncertainty. This reduces the maximum amplitude of the reference signal to be tracked depending on the maximum amplitude of uncertainty.*

Proof. From (5.23) and (5.24)

$$| F\tilde{x} + Hr + u_{N1} | \leq u_{max} \Rightarrow \bar{u} \leq u_{max}. \tag{5.26}$$

Therefore by design, the control law $\bar{u}$ does not saturate the input. Therefore from (5.1) and (5.8) the error dynamics can be written as follows

$$\dot{\tilde{x}} = (A+BF)\tilde{x} + Ax_e + BHr + \rho(x,t) + u_{N1}. \tag{5.27}$$

It is straight forward to see that $Ax_e + BHr = 0$. By using (5.17) the error dynamics becomes

$$\dot{\tilde{x}} = (A+BF)\tilde{x}, \tag{5.28}$$

which shows error dynamics is stable. Furthermore, it shows $X_\delta = \{\tilde{x}|\tilde{x}^T P\tilde{x} \leq c_\delta\}$ is an invariant set of closed loop system (5.28).

Remark 5.3.3. *In the above lemma to prove stability, linear controller is constrained to remain* $u_L \leq \varepsilon u_{max}$*. This will put an upper limit on the amplitude of* *the reference signal to be tracked. The following condition should satisfy to track a step signal of amplitude r*

$$|r| \leq \sqrt{c_\delta (G_e^T P G_e)^{-1}}, \tag{5.29}$$

$$|Hr| \leq \delta_1 u_{max}. \tag{5.30}$$

The inequality(5.29) can be proved as follows. From Lemma 1

$$\begin{aligned} x_e^T P x_e &\leq c_\delta, \\ \Rightarrow (G_e r)^T P (G_e r) &\leq c_\delta, \\ \Rightarrow r^T G_e^T P G_e r &\leq c_\delta, \\ \Rightarrow r^T r G_e^T P G_e &\leq c_\delta. \end{aligned}$$

By solving the above equation

$$|r| \leq \sqrt{c_\delta (G_e^T P G_e)^{-1}}.$$

Next, the closed loop error dynamics for system(5.1) with the input (5.14) is discussed.

From (5.1), (5.14) and (5.8) we can write the error dynamics as

$$\begin{aligned} \dot{\tilde{x}} &= (A+BF)\tilde{x} + Bw + B\rho(x,t) &(5.31)\\ &= (A+BF)\tilde{x} + Bg, &(5.32) \end{aligned}$$

where

$$\begin{aligned} g &= w + \rho(x,t), &(5.33)\\ w &= sat(F\tilde{x} + Hr + u_N + u_{N1}) - F\tilde{x} - Hr. &(5.34) \end{aligned}$$

u_{N1} will be replaced with its equivalent value from (5.17) for the analysis purpose as discussed in Section 3. In the following theorem stability of the closed loop system (5.1) with the control input (5.14) is proved.

Theorem 5.3.1. *Output of the system in (5.1) by the application of the control law proposed in (5.14) will track desired constant reference trajectory asymptotically when the conditions in (5.20), (5.22) and (5.23) are satisfied.*

Proof. For the error dynamics represented by (5.31), Let a Lyapunov be defined as

$$\begin{aligned} V &= \tilde{x}^T P \tilde{x}. &(5.35)\\ \Rightarrow \dot{V} &= -\tilde{x}^T W \tilde{x} + 2\tilde{x}^T P B g. &(5.36) \end{aligned}$$

Our aim is to ensure $\dot{V}$ should be negative definite. Here g is the only variable parameter. Now $\dot{V}$ is computed for three different cases.

Case 1: If

$$u = |F\tilde{x} + Hr + u_{N1} + u_N| \leq u_{max} \Rightarrow sat(u) = u,$$

therefore

$$sat(F\tilde{x} + Hr + u_{N1} + u_N) = F\tilde{x} + Hr + u_{N1} + u_N.$$

So

$$g = F\tilde{x} + Hr + u_{N1} + u_N - F\tilde{x} - Hr + \rho(x,t),$$

By using (5.17)

$$\begin{aligned} g &= u_N, \\ &= \Psi(r,y)B^T P\tilde{x}. \end{aligned}$$

It implies

$$\dot{V} = -\tilde{x}^T W\tilde{x} + 2\Psi(r,y)\tilde{x}^T PBB^T P\tilde{x} \leq -\tilde{x}^T W\tilde{x} \Rightarrow \dot{V} < 0.$$

Case 2: if

$$\begin{aligned} u &= F\tilde{x} + Hr + u_{N1} + u_N > u_{max} \\ &\text{therefore} \quad sat(F\tilde{x} + Hr + u_{N1} + u_N) = u_{max}. \end{aligned} \tag{5.37}$$

Therefore

$$g = u_{max} - F\tilde{x} - Hr - \rho(x,t).$$

By construction, $|Fx + Hr + \rho(x,t)| \leq u_{max}$, so $g > 0$ and from (5.37) and (5.17) $u_N > u_{max} - F\tilde{x} - Hr + u_{N1} > 0$. Which implies that $\tilde{x}^T PB < 0$. Hence

$$\tilde{x}^T PBg < 0.$$

This implies

$$\dot{V} = -\tilde{x}^T W\tilde{x} + 2\tilde{x}^T PBg \leq -\tilde{x}^T W\tilde{x} \Rightarrow \dot{V} < 0.$$

Case 3: If

$$\begin{aligned} u &= F\tilde{x} + Hr + u_{N1} + u_N < -u_{max} \\ &\text{therefore} \quad sat(F\tilde{x} + Hr + u_{N1} + u_N) = -u_{max}. \end{aligned} \tag{5.38}$$

So,

$$g = -u_{max} - F\tilde{x} - Hr + \rho(x,t)$$

By construction, $|F\tilde{x} + Hr + \rho(x,t)| \leq u_{max}$, so

$\Rightarrow g < 0$

From the above equation and using (5.38) and (5.17) it implies

$$u_N < -u_{max} - F\tilde{x} - Hr - u_{N1} < 0 \tag{5.39}$$

$\Rightarrow \tilde{x}^T PB > 0$

$\Rightarrow 2\tilde{x}^T PBg < 0.$

Therefore

$\dot{V} \leq -\tilde{x}^T W\tilde{x} \Rightarrow \dot{V} < 0.$

Therefore in all the cases when $\tilde{x}$ and r satisfy conditions (5.21) and (5.22) it has been shown that $\dot{V} < 0$ when $\tilde{x} \in X_\delta$ This implies

$$\lim_{t \longrightarrow \infty} x(t) = x_e. \tag{5.40}$$

Therefore

$$\lim_{t \longrightarrow \infty} h(t) = Cx_e = r. \tag{5.41}$$

The above theorem proves the stability of the closed loop system with the combined ISM-CNF law(5.14).

5.4 Experimental Evaluation for a Servo Position Control System

The servo system shown in Fig. 5.2 is an electromechanical system that represents important classes of systems such as conveyors, machine tools, spindle drives, and automated assembly machines. The setup is provided by Educational Control Products (ECP), California, USA. The system consists of a drive disc which is driven through a drive motor (servo actuator). The drive disc is coupled to the drive motor through a timing belt. The motion of the drive disc is transferred to another disc in which a load disc is used to load the system. The motion from drive disc to load disc is transferred through a speed reduction assembly and a timing belt. The load and the drive disc inertias are adjustable. High resolution encoder is used to measure the position of load disc. The drive motor is driven by a servo amplifier.

Following the procedure outlined in manual of Educational Control Product [32], the identified model for the system without any load on either discs can be given as follows

$$\begin{bmatrix} \dot{x_1} \\ \dot{x_2} \end{bmatrix} = \begin{bmatrix} 0 & 1 \\ 0 & -8.4344 \end{bmatrix} \begin{bmatrix} x_1 \\ x_2 \end{bmatrix} + \begin{bmatrix} 0 \\ 458.46 \end{bmatrix} (sat(u) + d) \tag{5.42a}$$

where x_1, x_2, u, and d are, respectively, the angular position of the load disc, the angular velocity of the load disc, input voltage to the drive motor, and disturbance voltage signal injected externally to perturb the plant. The maximum available input voltage is $u_{max} = 4.9\ Volts$. It should be noted that the above model captures a nominal dynamics of the plant. The purpose of giving an external disturbance is

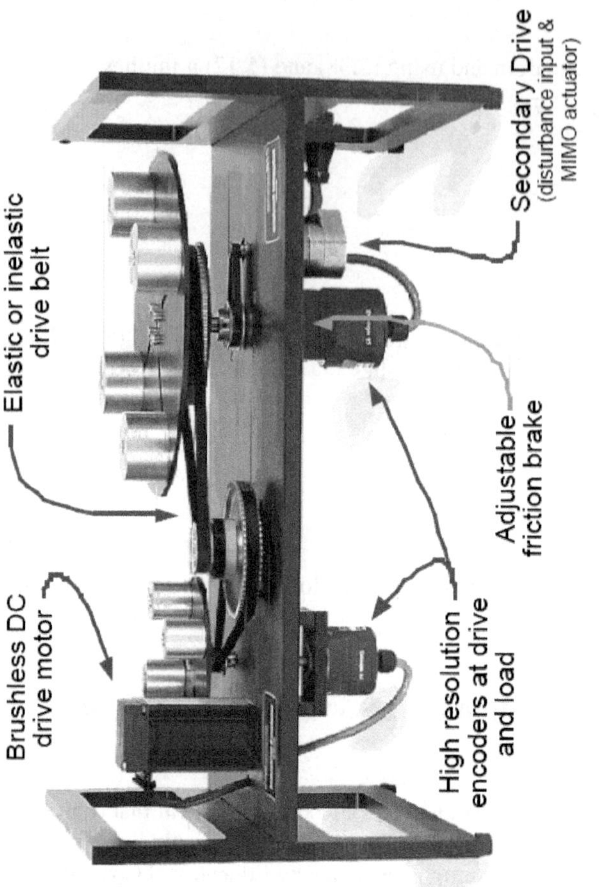

Fig. 5.2 Side view of laboratory based servo position control system

to check the robustness property of the system with the proposed algorithm. The disturbance term d is taken as follows

$$d = 0.2sin(20t). \tag{5.43}$$

It should be noted that in controller design only the bounds of this disturbance are used. What follows next is that a step by step procedure to design ISM-CNF controller is given.

Step 1: For settling time $t_s = 0.45$ *sec.* and damping ratio of $\zeta = 0.6$ the gain F becomes $F = [0.4787 \;\; 0.0204]$. The feedforward gain G_1 is computed from (5.6) as $G_1 = 0.4787$. Solve (5.19) for P with $W = 10^{-7}\begin{bmatrix} 0.25 & 0 \\ 0 & 0.25 \end{bmatrix}$ gives $P = 10^{-6}\begin{bmatrix} 0.1559 & 0.0001 \\ 0.0001 & 0.0007 \end{bmatrix}$ The above value of the matrix W is chosen after proper tun-

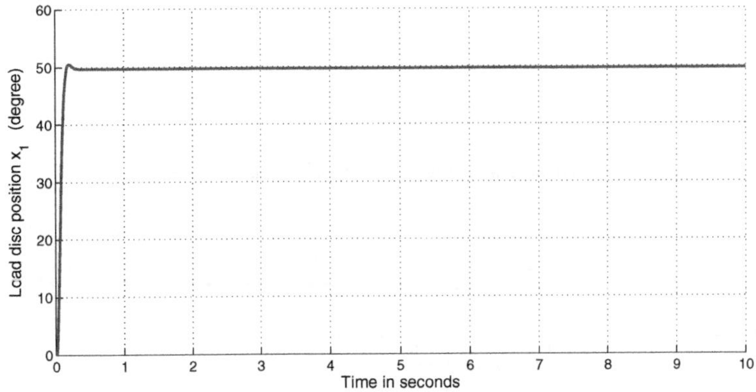

Fig. 5.3 Response of load disc to step signal of 50° in absence of any disturbance with CNF controller

ing. The nonlinear control part of CNF is designed as per (5.10) and (5.11) and with proper tuning one can obtain $\beta = -336$.

Step 2: To design u_{N1} as per (5.16) compute $G = B^+ = [0 \ \ 0.0022]$. The maximum value of the disturbance is $d_{max} = 0.2$, therefore M is chosen as 0.21. With these values control law (5.14) can be computed.

The above designed control law is implemented in real-time and experimentally tested for its performance in presence of sustained disturbance. The objective of the controller is to achieve desired angular position in face of disturbance. The reference step signal is to be tracked by the load disc is 50° for all the cases

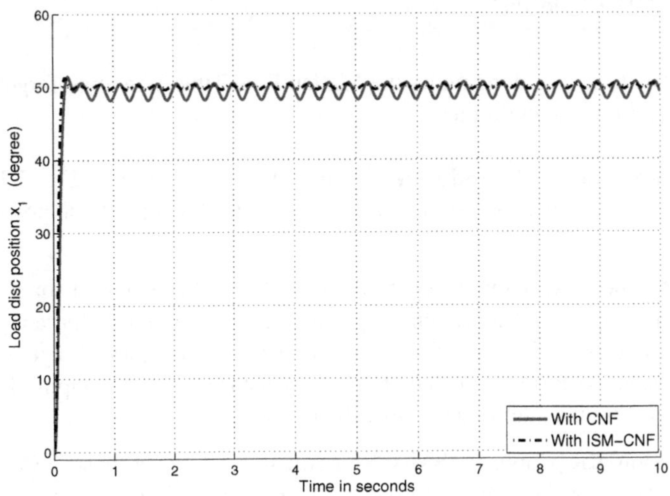

Fig. 5.4 Response of load disc position to step signal of 50° with disturbance; solid line with CNF and broken line with ISM-CNF

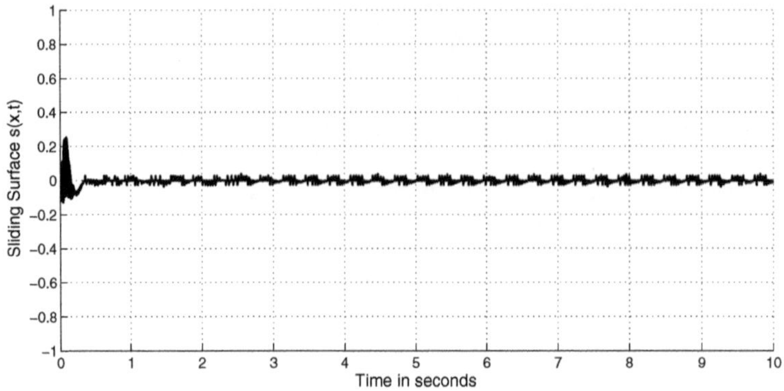

Fig. 5.5 Evolution of switching function s(x,t) with time

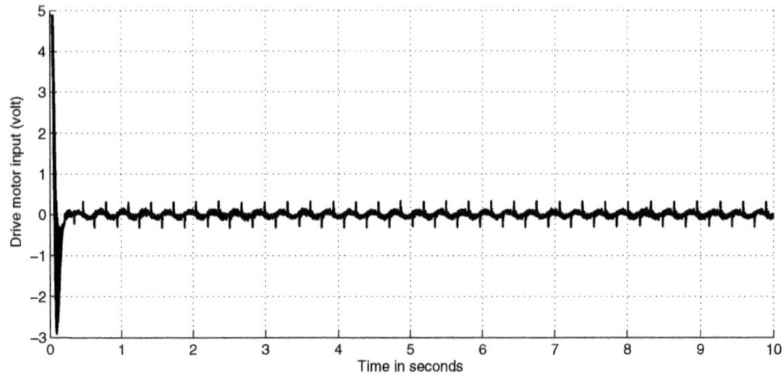

Fig. 5.6 Input with ISM-CNF algorithm

discussed below. The testing is done under the Matlab Real Time Windows Target (RTWT) under the following conditions:

Without disturbances (i.e $d = 0$) and CNF control: It can be seen from Fig. 5.3 that when there is no disturbance present in the system, the CNF control track position accurately.

With disturbances and CNF control: Now, the CNF controller is tested for its disturbance handling capacity. The system is perturbed by a sustain disturbance given by (5.43). It can be seen from Fig. 5.4 (solid line) which shows plot of angular position that the system performance is dictated by the disturbance. The amplitude of oscillation increases as the amplitude of disturbance is increased.

With disturbances and the proposed ISM-CNF control: Next, the proposed ISM-CNF controller is tested for the same disturbance. It can be seen from the Fig. 5.4 (broken line) that the controller takes counter action to suppress the effect of disturbance. Evolution of sliding surface with respect to time is shown in Fig. 5.5 which shows that $s(x,t)$ remains in the vicinity of zero from the beginning, therefore ensures the robustness in the entire motion. Fig. 5.6 shows plot of plot of input with ISM-CNF control law. After some initial transient, sinusoidal variations in input agrees to disturbance observer property of integral sliding mode(because disturbance is sinusoidally varying).

5.5 Conclusion

A new algorithm based on integral sliding mode and composite nonlinear control has been proposed. It has been shown that for uncertain system also how high performance feature of CNF can be retained by subtly combining it with the integral sliding mode. The same algorithm has been also evaluated through experiments on a laboratory based servo position control setup. The experimental results have shown significant improvement in the disturbance rejection property of the proposed ISM-CNF scheme.

Chapter 6
Multi-objective Sliding Mode Design Using Full-Order Lyapunov Matrix

6.1 Introduction

In this chapter, the sliding mode control (SMC) design for a class of continuous-time linear uncertain systems is considered based on the parametric approaches utilizing the Lyapunov (or Riccati-like) inequalities. SMC for the continuous-time linear systems is one of the well-known issues in control theory. However, this chapter addresses a new aspect of SMC for linear uncertain systems and establishes a systematic procedure to design a sliding hyperplane having multiple design objectives.

In the literature, much effort has been made to design sliding modes that satisfy the desired performance criteria. The well-known criteria include quadratic performance optimization [101], guaranteed H_2 cost minimization [90], eigen-structure assignment including pole-clustering [25, 31, 37], robustness to parametric uncertainties [88, 64, 107], and so on. Note that all these approaches are concerned with satisfying a single design objective. Moreover, the design objectives have not been presented in a unified framework.

On the other hand, remarkable progress has been made in linear control theory for solving the optimal problems with multiple constraints based on the LMIs (e.g., see [24] and [84]). The basic idea of the multi-objective approach based on LMIs is to seek a *common* Lyapunov matrix that simultaneously satisfies different parametric constraints imposed by the design performances. Assuming that the common variables may cause the conservatism, however, it does provide the flexibility of the control design with multiple objectives and the ease of synthesis in the parameter space based on the LMIs [17]. Also, there have been notable results in reducing the design conservatism in the literature (e.g., see [60] and [86]).

This chapter reconstructs the research results proposed in [61, 64, 62] and [63] in a unified LMIs framework. The proposed approach effectively solves the constrained optimization problems by adopting the common Lyapunov matrix idea of the linear control theory. First, through Sections 6.2 and 6.3, the sliding hyperplane design is considered for a class of uncertain systems with parametric uncertainties that are either *matched* or *mismatched*. Regardless of the matching condition of the model uncertainties, the quadratic stabilizability will be shown to

B. Bandyopadhyay et al.: Sliding Mode Control Using Novel Sliding Surfaces, LNCIS 392, pp. 97–113.
springerlink.com

be a sufficient condition for the existence of stable sliding modes. This enables the definition of a guaranteed cost optimization on the sliding hyperplane. Then, in Section 6.4, attention is paid to the quadratic performance optimization problem with the pole-clustering constraint in sliding mode. To this end, the parametric constraints for quadratic performance and pole-clustering requirement are derived in the forms of LMIs that are typically devised with the proposed parameterization technique, leading to the multi-objective optimization framework. Finally, in Section 6.5, the conclusion follows.

6.2 Parameterization of Sliding Mode Using the Lyapunov Matrix

6.2.1 System Description: Nominal Case

Consider the system

$$\dot{x}(t) = Ax(t) + Bu(t) + Fw(t), \tag{6.1}$$

where $x \in \mathbb{R}^n$ and $u \in \mathbb{R}^m$ are the state vector and the control input, respectively, and $w \in \mathbb{R}^l$ is a disturbance in which each element is bounded as $|w_j(t)| \leq \overline{w}_j$, $\forall j \in [0, l]$ for the known $\overline{w}_j$'s. Suppose that the system is of the regular form [101, 30]

$$\begin{cases} \dot{x}_1 = A_{11}x_1 + A_{12}x_2 \\ \dot{x}_2 = A_{21}x_1 + A_{22}x_2 + B_2u + F_2w, \end{cases} \tag{6.2}$$

where $x_1 \in \mathbb{R}^{n-m}$, $x_2 \in \mathbb{R}^m$, $B_2 \in \mathbb{R}^{m\times m}$ nonsingular and $F_2 \in \mathbb{R}^{m\times l}$. Observe that the matching condition is assumed for the external disturbance, $w(t)$. Without loss of generality, it can be seen that a similarity transformation always exists that converts the original system in a physical coordinate into the regular form system, as in (6.2). Also, let the stabilizability of the pair (A, B) be assumed.

Consider the switching function

$$s(t) = c_1x_1 + x_2 \tag{6.3}$$

for some $c_1 \in \mathbb{R}^{m\times(n-m)}$. Suppose that a control law is employed to satisfy the reachability condition such that $\dot{s}(t)^T s(t) < 0$, $\forall\, t > 0$, which achieves the sliding mode such that

$$s(t) = 0, \ \forall\, t \geq t_s$$

for some $t_s > 0$. In this chapter, let the control law be given by

$$u = \begin{cases} 0, & (\|s(t)\| = 0) \\ -B_2^{-1}\left\{c^TAx + \beta s + Z(t)sign(s)\right\}, & (\|s(t)\| > 0), \end{cases} \tag{6.4}$$

where $sign(s) = [sign(s_1), \cdots, sign(s_m)]^T$, $\beta > 0$, $c^T = [c_1,\ I_m]$, and $Z(t) = diag\,[z_1, \cdots, z_m]$ for z_i defined as

$$z_i = \sum_{k=1}^{l} |(c^T F)_{ik}| \overline{w}_k. \tag{6.5}$$

The detailed proof procedures are omitted since they are straightforward. Instead, the rough sketch for the proof is to show that

$$\dot{V}(s(t)) < -\beta \|s(t)\|^2$$

for a quadratic function, $V(s(t)) := \frac{1}{2} s(t)^T s(t)$, with the control law (6.4) [64].

Since $s(t) = 0$ in sliding mode, the dynamics of the system is confined in the reduced order space of $\mathbb{R}^{n-m}$, and the rest of the states in $\mathbb{R}^m$ are statically constrained as follows:

$$\begin{aligned} \dot{x}_1 &= (A_{11} - A_{12} c_1) x_1, \\ x_2 &= -c_1 x_1. \end{aligned} \tag{6.6}$$

Observe that a desired reduced order dynamics can be obtained by selecting the matrix c_1, and the reduce order state vector x_1 would converge to zero if the matrix $A_{11} - A_{12} c_1$ is stable. Also, the vector x_2 would converge to zero as does x_1.

6.2.2 All the Stabilizing Sliding Function Coefficients

The objective of this section is to introduce a unified means to parameterize all the feasible sliding hyperplanes by utilizing the full order Lyapunov matrices. Since the system stability is determined only by that of the reduced order dynamics in sliding mode, as demonstrated in (6.6), it may be necessary to handle the reduce order system (A_{11}, A_{12}) for designing stable sliding hyperplanes. This has been a standard concept for sliding mode design for a class of linear systems. However, it turns out that in [63] and [64] the original full order system (A, B) can be directly used, instead of (A_{11}, A_{12}), for obtaining a stable sliding surface by utilizing the full order Lyapunov matrices. Moreover, this idea significantly simplifies the synthesis of sliding hyperplanes because the resulting constraints are expressed by linear matrix inequalities (LMIs) [17].

Before stating the main result, a useful lemma is introduced for matrix manipulation in the following.

Lemma 1. *The matrix* $H := \begin{bmatrix} H_{11} & H_{12} \\ H_{12}^T & H_{22} \end{bmatrix}$ *is positive definite if and only if* $H_{22} > 0$ *and* $H_{11} - H_{12} H_{22}^{-1} H_{12}^T > 0$.

Proof. The proof is straightforward, as shown below:

$$\begin{bmatrix} 0 & I \\ I & -H_{12} H_{22}^{-1} \end{bmatrix} \begin{bmatrix} H_{11} & H_{12} \\ H_{12}^T & H_{22} \end{bmatrix} \begin{bmatrix} 0 & I \\ I & -H_{12} H_{22}^{-1} \end{bmatrix}^T = \begin{bmatrix} H_{22} & 0 \\ 0 & H_{11} - H_{12} H_{22}^{-1} H_{12}^T \end{bmatrix},$$

which completes the proof. □

Now, the concept for the parameterization of all the stable sliding hyperplanes can be presented in the following theorem.

Theorem 6.2.1. *Any stable sliding hyperplane exists in the form of*

$$c_1 = P_{22}^{-1} P_{12}^T \tag{6.7}$$

with any $0 < P \in \mathbb{R}^{n \times n}$ *satisfying, given a* $Q \geq 0$,

$$(A - BK)^T P + P(A - BK) + Q < 0 \tag{6.8}$$

for some $K \in \mathbb{R}^{m \times n}$, *where* P_{ij}*'s are defined by*

$$P = \begin{bmatrix} P_{11} & P_{12} \\ P_{12}^T & P_{22} \end{bmatrix} \in \begin{bmatrix} \mathbb{R}^{(n-m)\times(n-m)} & \mathbb{R}^{(n-m)\times m} \\ \mathbb{R}^{m\times(n-m)} & \mathbb{R}^{m\times m} \end{bmatrix}.$$

Proof. The proof is composed of two steps: to show that (i) a $P > 0$ exists satisfying (6.8) for any stabilizing c_1, and, (ii) for any $P > 0$ satisfying (6.8), the composition of c_1 as in (6.7) is the stabilizing sliding function coefficient.

Proof of (i): for an c_1 with the stability of $A_s := A_{11} - A_{12}c_1$, some $P_r > 0$ should exists satisfying

$$A_s^T P_r + P_r A_s + Q_r < 0, \tag{6.9}$$

where $Q_r := [I_{n-m} \; - c_1^T] Q [I_{n-m} \; - c_1^T]^T$ for any $Q \geq 0$. For later usage, let R denote $R := A_s^T P_r + P_r A_s + Q_r$, which is negative definite.

Now, for an arbitrary $P_{22} > 0$, define the matrices

$$P_{12} = c_1^T P_{22}, \;\; P_{11} = P_r + P_{12}^T P_{22}^{-1} P_{12}, \tag{6.10}$$

and

$$K = [K_1, \; K_2], \tag{6.11}$$

where, for an $\varepsilon > 0$,

$$\begin{aligned} K_1 &= B_2^{-1} \left\{ A_{21} + P_{22}^{-1} P_{12}^T A_{11} + P_{22}^{-1} A_{12}^T P_r + \tfrac{\varepsilon}{2} P_{22}^{-2} P_{12}^T \right\}, \\ K_2 &= B_2^{-1} \left\{ A_{22} + P_{22}^{-1} P_{12}^T A_{12} + \tfrac{\varepsilon}{2} P_{22}^{-1} \right\}. \end{aligned}$$

Using the matrices $P := \begin{bmatrix} P_{11} & P_{12} \\ P_{12}^T & P_{22} \end{bmatrix}$ and K defined above, it can be shown that, through some manipulations,

$$T \left\{ A_K^T P + P A_K + Q \right\} T^T = \begin{bmatrix} R & 0 \\ 0 & -\varepsilon I_m \end{bmatrix} < 0, \tag{6.12}$$

where $A_K = A - BK$ and $T = \begin{bmatrix} I_{n-m} & -P_{12}P_{22}^{-1} \\ 0 & I_m \end{bmatrix}$, which leads to (6.8).

Proof of (ii): define $T_r := \left[I_{n-m}, \; -P_{12}P_{22}^{-1} \right]$. Then, pre- and post multiplying (6.8) by T_r and T_r^T, respectively, yields

$$\left(A_{11} - A_{12} P_{22}^{-1} P_{12}^T \right)^T P_r + P_r \left(A_{11} - A_{12} P_{22}^{-1} P_{12}^T \right) + T_r Q T_r^T < 0, \tag{6.13}$$

where $P_r = P_{11} - P_{12}P_{22}^{-1}P_{12}^T$. Note that $P_r > 0$ by Lemma 1 since $P > 0$. Therefore, the matrix $A_{11} - A_{12}c_1$ is the stability matrix by the Lyapunov stability theorem with the choice of $c_1 = P_{22}^{-1}P_{12}^T$. This completes the proof. □

Theorem 6.2.1 shows that all the stable sliding hyperplanes can be explicitly expressed by combining two sub-partitions of the full order Lyapunov matrix (obtained by solving the Lyapunov inequality). As a matter of fact, the result by Theorem 6.2.1 unifies (and generalizes) the previous studies that utilized the Lyapunov equation or Riccati-like equations under some sufficient conditions (e.g., see [25, 64, 88]).

Additionally, observe that Theorem 6.2.1 presents the necessary and sufficient condition for the existence of stable sliding hyperplanes in terms of the solvability of the Lyapunov inequality. It is noted that the solvability of (6.8) is equivalent to the stabilizability of the pair (A,B). Therefore, it is apparent that the stabilizability of (A,B) also plays an important role for the existence of the sliding mode, which extends a certain controllability condition in [101].

6.3 Guaranteed Cost Sliding Mode Design for Uncertain Systems

In this section, a sliding mode design problem will be discussed for a class of uncertain systems with parametric uncertainties. To this end, the results stated in the previous section and [64] are extended and re-stated. Interestingly, it will be illustrated that the stable sliding modes exist under a quadratic stabilizability condition, even in the presence of mismatched uncertainties. Furthermore, the guaranteed cost control on the sliding hyperplane can be newly formulated.

Consider an uncertain system

$$\dot{x} = (A + \Delta A)x + Bu + Fw, \tag{6.14}$$

where ΔA represents the real parametric uncertainties of the form

$$\Delta A = MD(t)N, \tag{6.15}$$

where $M,\, N^T \in \mathbb{R}^{n\times h}$ and $D(t) = diag\left[\delta_1(t)I_{r_1}, \cdots, \delta_p(t)I_{r_p}\right]$ for the Lebesgue measurable functions δ_i such that $|\delta_i(t)| \leq 1,\ \forall\, t \geq 0$. The size of the identity matrix I_{r_i} is $r_i \times r_i$, where r_i is the repetition number of uncertain parameter δ_i in ΔA. See the example below to understand the description of parameter uncertainties. Also, the external disturbance is matched to the control input, and let the system be assumed to be in the regular form as follows: with invertible $B_2 \in \mathbb{R}^{m\times m}$ and $F_2 \in \mathbb{R}^{m\times l}$,

$$\begin{cases} \dot{x}_1 = (A_{11} + \Delta A_{11})x_1 + (A_{12} + \Delta A_{12})x_2 \\ \dot{x}_2 = (A_{21} + \Delta A_{21})x_1 + (A_{22} + \Delta A_{22})x_2 + B_2u + F_2w. \end{cases} \tag{6.16}$$

Example 1. Let ΔA be given in an additive form as follows:

$$\Delta A = \delta_1(t)E_1 + \delta_2(t)E_2 \tag{6.17}$$

with

$$E_1 = \begin{bmatrix} 4 & 0 \\ 0 & 0 \end{bmatrix}, \quad E_2 = \begin{bmatrix} 2 & 2 \\ 1 & 2 \end{bmatrix}.$$

Then, it can be seen that, with $r_1 = rank(E_1) = 1,\ r_2 = rank(E2) = 2$,

$$\begin{aligned}\Delta A &= \delta_1(t) \begin{bmatrix} 2 \\ 0 \end{bmatrix} \cdot \begin{bmatrix} 2 & 0 \end{bmatrix} + \delta_2(t) \begin{bmatrix} 0 & 2 \\ 1 & 1 \end{bmatrix} \cdot \begin{bmatrix} 0 & 1 \\ 1 & 1 \end{bmatrix} \\ &= \underbrace{\begin{bmatrix} 2 & 0 & 2 \\ 0 & 1 & 1 \end{bmatrix}}_{M} \underbrace{\begin{bmatrix} \delta_1(t) & & \\ & \delta_2(t) & \\ & & \delta_2(t) \end{bmatrix}}_{D(t)} \underbrace{\begin{bmatrix} 2 & 0 \\ 0 & 1 \\ 1 & 1 \end{bmatrix}}_{N}.\end{aligned}$$

The selection of M and N may not be unique; however, this would not be problematic because the scale matrices will be introduced later on, which eventually eliminate the dependency on the specific choices of M and N.

Now, suppose that the same control input (6.4), but, with

$$z_i = \sum_{j=1}^{h} |(c^T M)_{ij} (Nx)_j| + \sum_{k=1}^{l} |(c^T F)_{ik}| \overline{w}_k, \tag{6.18}$$

is applied to the uncertain system (6.14). Then, it is straightforward to show the reaching behavior by proving that $\dot{V} < -\beta \|s\|^2$ for a quadratic function $V = \frac{1}{2} s^T s$. The detailed procedures are omitted for saving space. Instead, [64] is referred to.

Now, to handle the robust stability to parameter uncertainties, let us introduce a sufficient condition for quadratic stability in the following.

Lemma 2. *Consider an uncertain system*

$$\dot{x} = (A_c + \Delta A)x, \tag{6.19}$$

where ΔA complies with (6.15). *Then, the system (6.19) is quadratically stable if there exist some $P > 0, X \in S_{sym}$ and $U \in S_{skew}$ satisfying,*

$$A_c^T P + PA_c + PMXM^T P + (N + PMU^T) X^{-1} (N + PMU^T)^T < 0, \tag{6.20}$$

where the sets for scales are defined as follows:

$$\begin{cases} S_{sym} := \{X \mid XD(t) = D(t)X,\ X > 0\} \\ S_{skew} := \{U \mid UD(t) = D(t)U,\ U = -U^T\} \end{cases}$$

Proof. For a quadratic function $V = x^T Px$, where $P > 0$, the quadratic stability is proven by showing $\dot{V} < 0$. To this end, note that

$$PMD(t)N + N^T D(t)^T M^T P = PMD(t)H + H^T D(t)^T M^T P \tag{6.21}$$

$$\leq PMXM^T P + H^T X^{-1} H, \tag{6.22}$$

where $H := N + UM^T P$. Note that (6.21) is established thanks to the commuting property and the skew symmetricity, i.e., $D(t)U = UD(t) = -U^T D(t)$. Also, (6.22) is a standard over-bounding technique for the block-diagonal uncertainties (e.g. see [64]). This completes the proof. □

The quadratic condition in (6.20) was first derived and applied to robust control problems in [85] and [67]. Note that the usage of the skew symmetric scale (as well as the symmetric scale) effectively reduces the design conservatism caused by the over-bounding process, in particular, in the presence of multiple (and multi-rank) uncertainties. See the example below for an illustration of the scale matrices.

Example 2. For the uncertainty $D(t)$ introduced in Example 1, the scale matrices are structured as follows: for scalars g_i's,

$$X = \begin{bmatrix} g_1 & 0 & 0 \\ 0 & g_2 & g_3 \\ 0 & g_3 & g_4 \end{bmatrix}, \; g_1 > 0, \; \begin{bmatrix} g_2 & g_3 \\ g_3 & g_4 \end{bmatrix} > 0, \; U = \begin{bmatrix} 0 & 0 & 0 \\ 0 & 0 & g_5 \\ 0 & -g_5 & 0 \end{bmatrix}.$$

Observe that the skew symmetric scale U is involved under the multi-rank uncertainty $\delta_2(t)$. In the absence of multi-rank uncertainties, the skew symmetric scale is obsolete. Also, note that the symmetric or skew symmetric scales can be easily incorporated as search variables with the LMIs solver of MATLAB.

Now, the guaranteed cost sliding mode design is introduced in the following theorem.

Theorem 6.3.1. *Sliding modes exist if some $P > 0$, $K \in \mathbb{R}^{m\times n}$, $X \in S_{sym}$ and $U \in S_{skew}$ exist satisfying*

$$\begin{aligned}(A-BK)^T P + P(A-BK) + Q + PMXM^T P \\ +(N+PMU^T)X^{-1}(N+PMU^T)^T < 0\end{aligned} \tag{6.23}$$

for matrix $Q \geq 0$. Moreover, the matrix $c_1 = P_{22}^{-1}P_{12}^T$, for any feasible matrix P, is the stabilizing sliding function coefficient and bounds the quadratic performance index as follows:

$$\int_{t_s}^{\infty} x^T Qx\, dt < x_1(t_s)^T P_r x_1(t_s), \tag{6.24}$$

where $P_r = P_{11} - P_{12}P_{22}^{-1}P_{12}^T$.

Proof. The proof procedures are similar to those proposed in Theorem 6.2.1. First, pre- and post multiply T_r and T_r^T by (6.23), respectively, where $T_r := \left[I_{n-m}, \; -P_{12}P_{22}^{-1}\right]$. Then, through some manipulations, one may show the quadratic stability of the reduced order uncertain system by choosing $c_1 = P_{22}^{-1}P_{12}^T$.

Secondly, to derive the upper bound of quadratic cost, consider a quadratic function: $V = x^T Px$ for $t \geq t_s$. Then, it is immediate to have

$$\dot{V} = x^T\left\{(A-BK+\Delta A)^T P + P(A-BK+\Delta A)\right\}x + 2x^T PB\left(u + B_2^{-1}F_2 w + Kx\right).$$

Note that $B^T P x = 0$ for $t \geq t_s$ since

$$\begin{aligned} s(t) &= x_2 + c_1 x_1 = P_{22}^{-1}\left[P_{12}^T,\, P_{22}\right] x \\ &= \left(P_{22}^{-1} B_2^{-T}\right) B^T P x, \end{aligned} \tag{6.25}$$

and $s(t) = 0$ for all $t \geq t_s$. Also, by applying the quadratic bounding technique with scales introduced in (6.22) to treat the uncertain terms, it can be shown that

$$\dot{V} < -x^T Q x,$$

which leads to

$$\begin{aligned} \int_{t_s}^{\infty} x^T Q x\, dt &< V(t_s) - V(\infty) \\ &\leq V(t_s) = x_1(t_s)^T \left(P_{11} - P_{12} P_{22}^{-1} P_{12}^T\right) x_1(t_s) \end{aligned} \tag{6.26}$$

because $x_2(t_s) = -P_{22}^{-1} P_{12}^T x_1(t_s)$ from the fact that $s = x_2 + P_{22}^{-1} P_{12}^T x_1 = 0$. This completes the proof. □

Considering that (6.23) would become equivalent to the Lyapunov inequality in (6.8) in the absence of uncertainties, i.e., $M = N = 0$, Theorem 6.3.1 can be viewed as an extension of Theorem 6.2.1 to a class of parametric uncertain systems. As far as the parameter uncertainty is concerned, the solvability of (6.23), which is a sufficient condition for quadratic stabilizability, is the existence condition for stable sliding hyperplanes, regardless of the *matching* condition. In the fields of SMC, it has been often assumed that the input-matching condition is required for achieving the robustness to uncertainties. However, from the result of Theorem 6.3.1, it is clear that the mismatched parameter uncertainties can be well treated by SMC, relying on the quadratic stability without the matching condition.

Another important aspect of Theorem 6.3.1 is that it provides a new concept for optimizing the quadratic cost index in sliding mode. Since the quadratic cost index is upper bounded by (6.24), the guaranteed cost sliding mode design can be defined by minimizing the upper bound of the quadratic performance. To this end, note that

$$Y = P^{-1} = \begin{bmatrix} Y_{11} & Y_{12} \\ Y_{12}^T & Y_{22} \end{bmatrix}, \tag{6.27}$$

where $Y_{ij}'s$ are given by

$$\begin{aligned} Y_{11} &= (P_{11} - P_{12} P_{22}^{-1} P_{12}^T)^{-1}, \\ Y_{12} &= -(P_{11} - P_{12} P_{22}^{-1} P_{12}^T)^{-1} P_{12} P_{22}^{-1}, \\ Y_{22} &= P_{22}^{-1} + P_{22}^{-1} P_{12}^T Y_{11} P_{12} P_{22}^{-1}, \end{aligned}$$

which shows that the upper bound in (6.24) can be expressed by Y_{11}. As a result, utilizing the change of variables and the Schur complement [17], the guaranteed cost sliding mode design can be defined by LMIs as presented in the following.

Guaranteed Cost Sliding Mode Design: Given a matrix $Q \geq 0$, minimize γ with respect to $Y > 0$, L, $X \in S_{sym}$ and $U \in S_{skew}$ satisfying

$$\begin{bmatrix} YA^T + AY - BL - L^T B^T + MXM^T & YC_q & YN + MU^T \\ C_q^T Y & -I & 0 \\ N^T Y + UM^T & 0 & -X \end{bmatrix} < 0, \tag{6.28}$$

and

$$\begin{bmatrix} \gamma & x_1(t_s)^T \\ x_1(t_s) & W_1 Y W_1^T \end{bmatrix} > 0, \tag{6.29}$$

where $Y = P^{-1}$, $L = KP^{-1}$, $Q = C_q C_q^T$, and $W_1 = [I_{n-m}, \ 0_{(n-m)\times m}]$. Then, for the optimal value of Y^*, the sliding function coefficient is given by

$$c_1^* = -Y_{12}^{*T} Y_{11}^{*-1}. \tag{6.30}$$

6.4 Multi-objective Sliding Mode Design

In the previous sections, it has been shown that the sliding hyperplane can be designed by utilizing the full order Lyapunov matrix. Furthermore, it should be noted that the features of the reduced order system (i.e., (A_{11}, A_{12})-system) can be exactly inherited from those of the full order system (i.e., (A,B)-system). This observation, in fact, leads to the further expansion of sliding mode design methods based on the multi-objective design. In this section, among many design criteria for selecting the sliding hyperplane, the quadratic performance optimization with a pole-clustering constraint is illustrated. To this end, mainly the results from [63] and [62] will be reconstructed.

6.4.1 Problem Statement

Suppose that the nominal system and the control input presented in Section 6.2.1 are adopted. Then, during the sliding mode phase, let the performance index be considered as follows: given a design parameter $0 < Q \in \mathbb{R}^{n\times n}$,

$$J(S) := \int_{t_s}^{\infty} x^T Q x \, dt \tag{6.31}$$

with the pole-clustering constraint

$$\lambda\,(A_{11} - A_{12}c_1) \subset \mathscr{L}(c,\rho), \tag{6.32}$$

where $\lambda(\cdot)$ is the set of eigenvalues of the argument matrix and $\mathscr{L}(c,\rho)$ is the circular region centered at $(-c,0)$ with the radius $\rho > 0$ and belonging to the left-half plane of the complex domain. The center c should be chosen such that $c > -\rho$ so that $\mathscr{L}(c,\rho)$ will be nonempty.

Observe that the problem above is composed of two design objectives: (i) quadratic performance optimization and (ii) the pole-clustering problem. To address these issues, efforts are made to derive the Lyapunov matrix based conditions first, and, then, the simultaneous synthesis with the derived conditions will be discussed in the subsequent subsections.

6.4.2 *Quadratic Performance Optimization*

The quadratic performance optimization in sliding mode was already treated in Section 6.3 in the presence of the parametric uncertainties. In fact, in the absence of the parameter uncertainties, the guaranteed cost sliding mode design introduced in Section 6.3 does become a quadratic cost minimization problem, which will be illustrated in an elaborate manner.

Given a matrix $Q > 0$, let the set be defined as follows.

$$\Omega(Q) := \left\{P \mid (A-BK)^T P + P(A-BK) + Q = 0,\ P > 0,\ K \in \mathbb{R}^{m\times n}\right\}, \quad (6.33)$$

which is nonempty with the stabilizability of the pair (A,B) (e.g., see [9]). Then, similarly to Theorem 6.2.1, the parameterization method can be introduced with the equality constraint in the following theorem.

Theorem 6.4.1. *Given any matrix $Q > 0$, all the stabilizing sliding function coefficients are given by*

$$c_1 = P_{22}^{-1} P_{12}^T \quad (6.34)$$

for some $P \in \Omega(Q)$, where P_{ij}'s are defined as

$$P = \begin{bmatrix} P_{11} & P_{12} \\ P_{12}^T & P_{22} \end{bmatrix} \in \begin{bmatrix} \mathbb{R}^{(n-m)\times(n-m)} & \mathbb{R}^{(n-m)\times m} \\ \mathbb{R}^{m\times(n-m)} & \mathbb{R}^{m\times m} \end{bmatrix}.$$

Also, for any $P \in \Omega(Q)$, the matrix $c_1 = P_{22}^{-1} P_{12}^T$ is the stabilizing sliding function coefficient.

Proof. Let c_1 be a stabilizing sliding function coefficient which guarantees the stability of the matrix $A_s := A_{11} - A_{12}c_1$. Define a positive definite matrix as $Q_r = \left[I_{n-m},\ -c_1^T\right] Q \left[I_{n-m},\ -c_1^T\right]^T$. Then, there should exist a $P_r > 0$ satisfying

$$(A_{11} - A_{12}c_1)^T P_r + P_r(A_{11} - A_{12}c_1) + Q_r = 0 \quad (6.35)$$

due to the stability of A_s. Now, for an arbitrary matrix $0 < P_{22} \in \mathbb{R}^{m\times m}$, define the matrices

$$P_{12} := c_1^T P_{22},\ P_{11} := P_r + P_{12}^T P_{22}^{-1} P_{12},\ K := [K_1,\ K_2], \quad (6.36)$$

where

$$K_1 = B_2^{-1}\left\{A_{21} + P_{22}^{-1} P_{12}^T A_{11} + P_{22}^{-1} A_{12}^T P_r\right\},\ K_2 = B_2^{-1}\left\{A_{22} + P_{22}^{-1} P_{12}^T A_{12}\right\}.$$

Note that $P := \begin{bmatrix} P_{11} & P_{12} \\ P_{12}^T & P_{22} \end{bmatrix}$ is positive definite since $P_{22} > 0$ and $P_{11} - P_{12}P_{22}^{-1}P_{12}^T$ $(= P_r) > 0$. With the matrices P and K above, it can be shown that

$$\begin{bmatrix} I_{n-m} & -c_1^T \\ 0 & I_m \end{bmatrix} \{(A-BK)^T P + P(A-BK) + Q\} \begin{bmatrix} I_{n-m} & -c_1^T \\ 0 & I_m \end{bmatrix}^T = 0, \tag{6.37}$$

which implies that $P \in \Omega(Q)$.

Now, for any $P \in \Omega(Q)$, define as $T_r := \left[I_{n-m}, \; -P_{12}P_{22}^{-1}\right]$. Then, pre- and post-multiplying the Lyapunov equation in (6.33) by T_r and T_r^T, respectively, yields

$$\left(A_{11} - A_{12}P_{22}^{-1}P_{12}^T\right)^T P_r + P_r\left(A_{11} - A_{12}P_{22}^{-1}P_{12}^T\right) + T_r Q T_r^T = 0, \tag{6.38}$$

where $P_r = P_{11} - P_{12}^T P_{22}^{-1} P_{12}$, which is positive definite since $P > 0$. Then, by choosing $c_1 = P_{22}^{-1}P_{12}^T$, the asymptotic stability of $A_{11} - A_{12}c_1$ is guaranteed, which completes the proof. □

Theorem 6.4.1 is equivalent to Theorem 6.2.1 except for the constraint style adopted. It is noted that the matrix Q does not limit the feasible solution space, but specifies a certain performance index, which will become clearer in the following theorem.

Theorem 6.4.2. *Given a stabilizing sliding function coefficient c_1, the quadratic cost function (6.31) is expressed as follows:*

$$J(c_1) = x_1(t_s)^T P_r x_1(t_s) \tag{6.39}$$

for $P \in \Omega(Q)$ satisfying (6.34), where $P_r = P_{11} - P_{12}P_{22}^{-1}P_{12}^T$.

Proof. For the sliding function coefficient c_1, a $P \in \Omega(Q)$ exists satisfying $c_1 = P_{22}^{-1}P_{12}^T$ for some K (by Theorem 6.4.1). Then, it follows that

$$s(t) = x_2 + c_1 x_1 = P_{22}^{-1}\left[P_{12}^T \; P_{22}\right] x = \left(P_{22}^{-1}B_2^{-T}\right) B^T P x,$$

which implies $B^T P x = 0$ when $s(t) = 0$ for $t \ge t_s$ with nonsingular $B_2^T P_{22}$. Note that $B^T = \left[0_{m\times(n-m)}, \; B_2^T\right]$ from (6.1) and (6.2).

To calculate the derivative of a quadratic function $V = x^T P x$ for $t \ge t_s$, use the system equation rewritten by

$$\dot{x} = (A-BK)x + B(u + Dw + Kx)$$

for the matrix K associated with the matrix P. It leads to

$$\begin{aligned} \dot{V} &= x^T\left\{(A-BK)^T P + P(A-BK)\right\}x + 2x^T PB(u + Dw + Kx) \\ &= -x^T Q x + 2x^T PB(u + Dw + Kx) = -x^T Q x \end{aligned} \tag{6.40}$$

since $B^T P x = 0$ on $s(t) = 0$. Integrating both sides, it holds that

$$\int_{t_s}^{\infty} x^T Qx\, dt = x(t_s)^T Px(t_s) = x_1(t_s)^T \left(P_{11} - P_{12}P_{22}^{-1}P_{12}^T\right) x_1(t_s)$$

since $x_2(t_s) = -P_{22}^{-1}P_{12}^T x_1(t_s)$. This completes the proof. □

As a matter of fact, the results from Theorems 6.4.1 and 6.4.2 prove that

$$\min_{c_1} J(c_1) = \min_{P \in \Omega(Q)} x_1(t_s)^T P_r x_1(t_s), \tag{6.41}$$

which allows the convex optimization in the parameter space of (P,K) with the Lyapunov equation based on LMIs techniques [17].

Using the change of variables such as $Y := P^{-1}$ and $L := KP^{-1}$ and the relationship in (6.27), it may be straightforward to define the linear quadratic sliding mode design as follows.

Linear Quadratic Sliding Mode Design: Given a $Q > 0$, minimize γ with respect to $Y > 0$ and L satisfying

$$\begin{bmatrix} AY + YA^T - BL - L^T B^T & YC_Q \\ C_Q^T Y & -I_q \end{bmatrix} < 0 \tag{6.42}$$

$$\begin{bmatrix} \gamma I_{n-m} & I_{n-m} \\ I_{n-m} & Y_{11} \end{bmatrix} > 0, \tag{6.43}$$

where $Q = C_Q C_Q^T$ and $q = rank(Q)$. Then, for the optimal value of Y^*, the sliding function coefficient is given by

$$c_1^* = -Y_{12}^{*T} Y_{11}^{*-1}. \tag{6.44}$$

Remark 1. Note that the inequality (6.43) would introduce the design conservatism to some extent but remove the dependency on the initial state that is hardly known in practice. Thus, the linear quadratic sliding mode design, in fact, optimizes the upper bound of the quadratic performance index such that

$$\int_{t_s}^{\infty} x^T Qxdt < \gamma \|x_1(t_s)\|^2.$$

Regarding the quadratic performance optimization, the standard approach in SMC design has been to use the algebraic Riccati equation for the reduced order system, i.e., (A_{11}, A_{12})-system [101, 36]. Compared to the conventional approach, the proposed method handles the full order system so that the solution description becomes remarkably simplified and, moreover, the issues on the existence of a sliding hyperplane and the optimality are clearly stated.

6.4.3 Pole-Clustering Problem in Sliding Mode

In this section, a method is addressed to represent the pole-clustering constraint (6.32) in terms of the parametric approach. To begin with, a result derived from the literature [108, 24] is introduced in the following lemma.

Lemma 3. *The pole clustering constraint (6.32) is met by the matrix* c_1 *if and only if some* $0 < P_r \in \mathbb{R}^{(n-m)\times(n-m)}$ *exist satisfying*

$$\begin{bmatrix} \rho P_r & (A_{11}-A_{12}c_1)^T P_r + cP_r \\ P_r(A_{11}-A_{12}c_1)+cP_r & \rho P_r \end{bmatrix} > 0. \tag{6.45}$$

Based on the lesson from the previous sections, it may be noted that the Lyapunov matrix P_r of reduced order would be related to that of full order. The answer to this intrinsic argument can be presented in the following theorem.

Theorem 6.4.3. *Given any sliding function coefficient* c_1 *satisfying (6.32), there exist some* $0 < P \in \mathbb{R}^{n\times n}$ *that admit* $c_1 = P_{22}^{-1}P_{12}^T$ *and*

$$\begin{bmatrix} \rho P & (A-BK)^T P + cP \\ P(A-BK)+cP & \rho P \end{bmatrix} > 0 \tag{6.46}$$

for some $K \in \mathbb{R}^{m\times n}$*. Moreover, For any* $P > 0$ *satisfying (6.46), the matrix given by* $c_1 := P_{22}^{-1}P_{12}^T$ *meets (6.32).*

Proof. Let c_1 be the sliding function coefficient matrix satisfying (6.32). Then, from the necessity of Lemma 3, a $P_r > 0$ exists satisfying (6.45). Define the matrices for an arbitrary $0 < P_{22} \in \mathbb{R}^{m\times m}$ as

$$P_{12} := c_1^T P_{22}, \;\; P_{11} := P_r + P_{12}^T P_{22}^{-1} P_{12}, \;\; K := [K_1,\, K_2], \tag{6.47}$$

where

$$\begin{aligned} K_1 &= B_2^{-1}\left\{P_{22}^{-1}P_{12}^T(A_{11}+cI_{n-m})+A_{21}\right\}, \\ K_2 &= B_2^{-1}\left\{P_{22}^{-1}P_{12}^T A_{12}+A_{22}+cI_m\right\}. \end{aligned}$$

Note that the matrix $P := \begin{bmatrix} P_{11} & P_{12} \\ P_{12}^T & P_{22} \end{bmatrix}$ is positive definite since $P_{22} > 0$ and $P_r > 0$. Now, it will be shown that P and K satisfy (6.46). To this end, consider the matrices

$$H = \left[\begin{array}{cc|cc} 0_{m\times(n-m)} & I_m & 0_{m\times(n-m)} & 0_m \\ 0_{m\times(n-m)} & 0_m & 0_{m\times(n-m)} & I_m \end{array}\right]$$

and

$$T = \left[\begin{array}{cc|cc} I_{n-m} & -P_{12}P_{22}^{-1} & 0_{n-m} & 0_{(n-m)\times m} \\ \hline 0_{n-m} & 0_{(n-m)\times m} & I_{n-m} & -P_{12}P_{22}^{-1} \end{array}\right],$$

which have the matrix $\begin{bmatrix} H \\ T \end{bmatrix} \in \mathbb{R}^{2n\times 2n}$ nonsingular. Let $L_{(6.45)}$ and $L_{(6.46)}$ denote the left-hand sides of (6.45) and (6.46), respectively. Some elaborate manipulations lead to

$$\begin{bmatrix} H \\ T \end{bmatrix} L_{(6.46)} \begin{bmatrix} H \\ T \end{bmatrix}^T = \left[\begin{array}{c|c} HL_{(6.46)}H^T & HL_{(6.46)}T^T \\ \hline TL_{(6.46)}H^T & TL_{(6.46)}T^T \end{array}\right]$$
$$= \left[\begin{array}{cc|c} \rho P_{22} & 0 & \\ 0 & \rho P_{22} & 0 \\ \hline \multicolumn{2}{c|}{0} & L_{(6.45)} \end{array}\right],$$

which proves the inequality of (6.46) since $P_{22} > 0$ and $L_{(6.45)} > 0$.

Now, to prove the second argument, let an augmented matrix be defined as

$$T = \begin{bmatrix} T_r & 0_{(n-m)\times n} \\ 0_{(n-m)\times n} & T_r \end{bmatrix},$$

where T_r is given by (6.38). By pre- and post-multiplying (6.46) by T and T^T, respectively, it can be shown that (6.45) holds with the selection of $c_1 = P_{22}^{-1}P_{12}^T$ and $P_r = P_{11} - P_{12}^T P_{22}^{-1} P_{12}$. This completes the proof. □

Note that the inequality (6.46) is for placing the closed loop eigenvalues in the specific region such that $\lambda(A - BK) \subset \mathscr{D}(c,\rho)$ by the full state feedback $u = -Kx$. Therefore, as in the previous sections, it can be seen that the pole-clustering property in the reduced order system can be inherited from that of the full order system.

Sliding Mode Design with Pole-Clustering Constraint: Suppose that a region $\mathscr{D}(c,\rho)$ is given in the complex domain. Find $Y = Y^T > 0$ and $L \in \mathbb{R}^{m\times n}$ satisfying

$$\begin{bmatrix} \rho Y & YA^T - L^T B^T + cY \\ AY - BL + cY & \rho Y \end{bmatrix} > 0. \tag{6.48}$$

Then, the sliding function coefficient that places the eigenvalues of the sliding mode in the region is given by $c_1 = -Y_{12}^T Y_{11}^{-1}$.

6.4.4 Convex Synthesis for Multi-objective Design

From the results in Sections 6.4.2 and 6.4.3, it may be seen that the sliding function coefficients can be generated by the same composition structure (i.e., $c_1 = -Y_{12}^T Y_{11}^{-1}$) but with different Lyapunov matrices defining different design objectives (i.e., one for quadratic performance and the other for pole-clustering). For a multi-objective design, as addressed in Section 6.4.1, observe that two design objectives can be simultaneously met by the single sliding mode if

$$Y_{LQ,12}^T Y_{LQ,11}^{-1} = Y_{PL,12}^T Y_{PL,11}^{-1}, \tag{6.49}$$

where Y_{LQ} and Y_{PL} denote the inverse of Lyapunov matrices for the quadratic performance optimization in (6.42) and (6.43) and the pole-clustering problem in (6.48), respectively. Note that the variables Y_{LQ} and Y_{PL} are convex in the inequality constraints, but, not jointly with (6.49).

To resolve the nonconvexity, there have been several approaches in the literature for LMIs [24, 60, 86]. However, the problem still remains unsolved. One of the easiest methods to eliminate the nonconvexity in (6.49) is to assume that $Y_{LQ} = Y_{PL}$. However, this would lead to overly conservative results such as no solution found or an overly high upper bound for the quadratic performance index.

To derive a convex formula which is less conservative, let the Lyapunov matrices be composed as

$$Y_{LQ} = \begin{bmatrix} Y_{11} & Y_{12} \\ Y_{12}^T & Z_1 \end{bmatrix}, \quad Y_{PL} = \begin{bmatrix} Y_{11} & Y_{12} \\ Y_{12}^T & Z_2 \end{bmatrix} \tag{6.50}$$

for some Y_{11}, Y_{12}, Z_1 and Z_2 with the dimensions as partitioned in (6.34), which satisfies (6.49). Therefore, the multi-objective sliding mode design can be described with convexity in the following theorem.

Theorem 6.4.4. *Given a $Q > 0$ and $\mathscr{L}(c,\rho)$, the sliding function coefficient c_1 satisfying*

- $\lambda(A_{11} - A_{12}c_1) \subset \mathscr{L}(c,\rho)$*, and,*
- $\int_{t_s}^{\infty} x^T Q x dt < \gamma \|x_1(t_s)\|^2$

exists if some Y_{11}, Y_{12}, Z_1, Z_2, L_1, L_2, and γ exist satisfying

$$Y_{LQ} > 0 \tag{6.51}$$

$$Y_{PL} > 0 \tag{6.52}$$

$$\begin{bmatrix} \rho Y_{PL} & Y_{PL}A^T - L_1^T B^T + cY_{PL} \\ AY_{PL} - BL_1 + cY_{PL} & \rho Y_{PL} \end{bmatrix} > 0 \tag{6.53}$$

$$\begin{bmatrix} AY_{LQ} + Y_{LQ}A^T - BL_2 - L_2^T B^T & Y_{LQ}C_Q \\ C_Q^T Y_{LQ} & -I_q \end{bmatrix} < 0 \tag{6.54}$$

$$\begin{bmatrix} \gamma I_{n-m} & I_{n-m} \\ I_{n-m} & Y_{11} \end{bmatrix} > 0 \tag{6.55}$$

where Y_{LQ} and Y_{PL} are defined in (6.50). Also, the admissible sliding function coefficient is given by $c_1 = -Y_{12}^T Y_{11}^{-1}$ for the feasible parameters in (6.51)–(6.55).

It is noted that all the search variables are linear in the inequality constraints. By optimizing γ with the LMIs numerical tools, the linear quadratic sliding mode with the pole-clustering constraint can be easily obtained while a certain conservatism may exist. However, considering that the convexity in Theorem (6.4.4) allows a remarkable convenience for the multi-objective sliding mode design in a numerical aspect, Theorem 6.4.4 is worth applying to obtain an admissible design. See the design example below.

The idea demonstrated in this section can be further applied to several other design objectives. For example, the guaranteed cost sliding mode design in Section 6.3 can be combined with the robustness constraint to the time delay in the system states [62].

Example 3. Consider the system addressed in [88] as follows:

$$A = \begin{bmatrix} 0.2325 & -0.9285 & 0.0154 & 0.1222 \\ -0.7274 & 1.0116 & -0.0224 & 0.1576 \\ -1.6883 & 0.2214 & 0.6534 & 1.6278 \\ -0.5310 & -0.2603 & -0.0052 & 1.1025 \end{bmatrix}, \quad B = \begin{bmatrix} 3.00 & 2.00 \\ 0 & 1.00 \\ 0.50 & -2.00 \\ 1.30 & 0 \end{bmatrix}.$$

Let the design objectives be selected as

$$Q = I_{4\times4}, \quad c = 2, \quad \rho = 0.5.$$

To obtain the regular form in (6.2), a state transformation matrix is set as $T = [U_2,\ U_1]^T$, where U_1 and U_2 are obtained by the singular value decomposition of B such that $B = [U_1,\ U_2] \begin{bmatrix} \sigma \\ 0_{2\times2} \end{bmatrix} V^T$. Then, we can start the design procedure using the transformed system in a regular form:

$$A := TAT^{-1}, \quad B := TB, \quad Q := T^{-T}QT^{-1}.$$

Based on the LMIs representations in Theorem 6.4.4, the LMIs solver (i.e., LMIs Toolbox in MATLAB) may produce the minimal γ and the associated sliding function coefficient matrix as follows:

$$\gamma^* = 427.44, \quad c_1^* = \begin{bmatrix} -2.8104 & -1.274 \\ -3.8506 & 36.686 \end{bmatrix}.$$

It can be seen that the eigenvalues of the sliding mode are placed at $\{-1.6244 \pm 0.24845i\}$, which are all inside the constraint circle.

6.4.5 *Further Discussions: Nonconvex Synthesis for Multi-objective Design*

Reducing the conservatism in the convex formula may be of concern in order to obtain a less conservative design regardless of the numerical difficulty. To this end, instead of (6.50), let the Lyapunov matrices be composed as follows:

$$Y_{LQ} = \begin{bmatrix} Z_0 & Z_0 Y_{11}^{-1} Y_{12} \\ Y_{12}^T Y_{11}^{-1} Z_0 & Z_1 \end{bmatrix}, \quad Y_{PL} = \begin{bmatrix} Y_{11} & Y_{12} \\ Y_{12}^T & Z_2 \end{bmatrix} \tag{6.56}$$

for some Z_0, Z_1, Y_{11}, Y_{12} and Z_2. This structure can be assumed without loss of generality to necessarily satisfy (6.49) and results in a single sliding hyperplane for two design objectives. Note that, if $Z_0 = Y_{11}$, the convexity can be restored as stated in Section 6.4.4.

Under the structures proposed in (6.56), it is noted that the convexity for two inequalities (6.51) and (6.54) would be destroyed among the inequalities (6.51)–(6.55). So far, this problem remains unsolved. However, fortunately, an effective numerical algorithm has been proposed in [63] that is iterative and successively feasible with LMIs. Starting with a set of feasible parameters, it was shown that the cost index successively decreases and the LMIs are guaranteed to be feasible at every iteration. Also, it was illustrated that the initially feasible parameters can be always found either by solving Theorem 6.4.4 or by a reconstruction method if Theorem 6.4.4 fails to yield a solution. The numerical algorithm is based on the bounding techniques with generalized scales introduced in [86]. Regarding the detailed procedures for the numerical algorithm, [63] is referred to. It may be seen that, for Example 3, the numerical algorithm in [63] results in

$$\gamma^* = 313.56, \;\; c_1^* = \begin{bmatrix} -1.6732 & -15.951 \\ -2.3586 & 21.139 \end{bmatrix},$$

which locates the sliding mode poles at $s = -1.5$, which is exactly on the constraint circle boundary.

6.5 Conclusion

In this chapter, a unified framework of sliding mode design for a class of linear time-invariant continuous systems was proposed based on the parametric approach in which the full order Lyapunov matrix is used for representing the feasible sliding hyperplanes. For imposing the desired design objective in the sliding mode, it was shown that the LMIs developed for full state feedback can be adopted for the sliding mode design without loss of generality. The topics mainly handled in this chapter includes the parameterization of all the stabilizing sliding hyperplane for the nominal system, the linear quadratic sliding mode design, the guaranteed cost control for uncertain systems, the pole-clustering in sliding mode and the multi-objective sliding mode design. All these topics provide new insights for interpreting the sliding mode control, and, the systematic methods to design the sliding hyperplanes, in particular, with multiple design objectives.

Chapter 7
Lyapunov-Based Sliding Mode Control with Multi-Rate Output Feedback

7.1 Introduction

In the preceding chapter a design method for sliding surface is presented for continuous time system using full order Lyapunov matrix. Also this method is extended for to continuous uncertain time system. It is also noted as elaborated in introduction chapter that discrete-time sliding mode control is gaining more necessity since the actual implementation of control is generally carried out with digital signal processors. Recently, efforts have been made to recover the robustness of DSMC such as the discrete approximation approach [102], the sliding sector [45], and the quasi-sliding modes [92, 12]. Here, it is pointed out that the research on the sliding hyperplane design itself has drawn little attention in the literature relative to the studies on reaching law design. As for the sliding hyperplane design, the eigenvalue assignment methods for the equivalent dynamics matrix have been considered as standard (e.g., see Tang & Misawa [92] and references therein). Apart from those eigenvalue approaches, the parametric approaches utilizing the Lyapunov equation or the Riccati equation have been investigated in some studies. Spurgeon [87] introduced the usage of the Lyapunov matrix for calculating the sliding hyperplanes. In [45], a Riccati equation is used for the discrete-time sliding mode design for a class of single input systems. Recently, the LQ optimization approach was proposed by Janardhanan & Kariwala [55].

In this chapter, a whole sequence of design steps for DSMC is proposed based on the recent results in the literature [59, 91, 53]. First, in Section 7.2, the Lyapunov matrix approach treated in Chapter 6 is extended for the discrete-time sliding hyperplane design. The proposed approach will show the necessary and sufficient conditions for the existence of stable sliding hyperplanes. Also, all the stable sliding hyperplane can be parameterized with the full order Lyapunov matrix. Then, in Section 7.3, a reaching law is derived, which achieves "quasi-sliding" mode by adopting the saturation function instead of the signum function. To admit a reaching law with the partial measurements, in Section 7.4, the multi-rate output feedback is adopted. This shows that the sliding function can reside within an extended bound

B. Bandyopadhyay et al.: Sliding Mode Control Using Novel Sliding Surfaces, LNCIS 392, pp. 115–129.
springerlink.com

only with the partial measurements. Finally, in Section 7.5, the proposed design scheme is validated with an example by simulation.

7.2 Discrete-Time Sliding Hyperplane Design

This section focuses on the derivation of all the discrete-time sliding hyperplanes on which the finite-time convergence is assured, based on the discrete-time Lyapunov inequality.

Before stating the main results, it is necessary to address the discrete-time stabilizability condition by means of the discrete-time Lyapunov inequality. To this end, let a discrete-time LTI system be described, with appropriate dimensions, by

$$x_s(k+1) = \Phi_s x_s(k) + \Gamma_s u_s(k), \tag{7.1}$$

where $x_s(k)$ and $u_s(k)$ are the state and the control input, respectively. Then, the stabilizability via full state feedback can be expressed by the discrete-time Lyapunov inequality in the following.

Theorem 7.2.1. *Given a matrix $Q_s \geq 0$, the system (7.1) is stabilizable by a full state feedback $u_s(k) = -Kx_s(k)$ if and only if matrices $P_s > 0$ and K exist satisfying*

$$(\Phi_s - \Gamma_s K)^T P_s (\Phi_s - \Gamma_s K) - P_s + Q_s < 0. \tag{7.2}$$

It is noted that the solvability of the discrete-time Lyapunov inequality depends on not the selection of the matrix Q_s but the stabilizability of the pair (Φ_s, Γ_s). Moreover, the matrix $\Phi_s - \Gamma_s K$ with the resulting matrix K should be asymptotically stable, i.e., all the eigenvalues of the matrix resides inside the unit circle in the complex domain.

Now, let the system of interest be given by

$$x(k+1) = \Phi x(k) + \Gamma u(k) + F w(k), \tag{7.3}$$

where $x(k) \in \mathbb{R}^n$, $u(k) \in \mathbb{R}^m$ and $w(k) \in \mathbb{R}^l$ are the state variable, the control input, and the disturbance, respectively. The disturbance is assumed to be bounded as follows:

$$|w_j(k)| < \overline{w}_j, \ \forall\, j = 1, \cdots, l.$$

Furthermore, for simplicity, let the system be assumed to be in regular form as follows:

$$\Gamma = \begin{bmatrix} 0_{(n-m)\times m} \\ \Gamma_2 \end{bmatrix}, \ F = \begin{bmatrix} 0_{(n-m)\times l} \\ F_2 \end{bmatrix}$$

and

$$\Phi = \begin{bmatrix} \Phi_{11} & \Phi_{12} \\ \Phi_{21} & \Phi_{22} \end{bmatrix}$$

for $\Gamma_2 \in \mathbb{R}^{m\times m}$ invertible and $F_2 \in \mathbb{R}^{m\times l}$. Note that, without loss of generality, it is always possible to transform the system in general coordinates into the regular form system by properly choosing the coordinate transformation. For example, using the singular value decomposition of Γ such that $\Gamma = U_1 \Sigma V^T$ for the unitary matrix $[U_1,\, U_2] \in \mathbb{R}^{n\times n}$, a state transformation such as $x' = Tx$, leads to the regular form system, where $T = [U_2,\, U_1]^T$. Note that the disturbance exists with the control inputs, which implies the so-called *matching* condition.

Define the sliding function as

$$s(k) = x_2(k) + c_1 x_1(k). \tag{7.4}$$

Then, when an appropriate control input satisfying the *reaching law* is applied, the sliding behavior is expressed by

$$x_1(k+1) = (\Phi_{11} - \Phi_{12} c_1) x_1(k), \tag{7.5}$$

$$x_2(k) = -c_1 x_1(k). \tag{7.6}$$

This shows that the coefficient matrix c_1 of the sliding function has the role of the state feedback gain matrix for the reduced order system defined by (Φ_{11}, Φ_{12}). It is noted that all the feasible matrices for c_1 can be parameterized by the discrete-time Lyapunov inequality for the original full order system as illustrated below.

Theorem 7.2.2. *Given a matrix $Q \geq 0$, let us define a set of matrices as follows:*

$$\Omega(Q) := \{P > 0 | (\Phi - \Gamma K)^T P (\Phi - \Gamma K) - P + Q < 0\}. \tag{7.7}$$

Then, the matrix defined as

$$c_1 := P_{22}^{-1} P_{12}^T \tag{7.8}$$

for any matrix $P \in \Omega(Q)$, is the stable sliding function coefficient, where

$$P := \begin{bmatrix} P_{11} & P_{12} \\ P_{12}^T & P_{22} \end{bmatrix} \in \begin{bmatrix} \mathbb{R}^{(n-m)\times(n-m)} & \mathbb{R}^{(n-m)\times m} \\ \mathbb{R}^{m\times(n-m)} & \mathbb{R}^{m\times m} \end{bmatrix}.$$

Moreover, for any stable sliding function coefficient matrix c_1, a matrix $P \in \Omega(Q)$ exists satisfying (7.7) *and* (7.8).

Proof. (1st argument): For any $P \in \Omega(Q)$, it can be shown that

$$\begin{aligned}
&\Phi^T P \Phi - \Phi^T P \Gamma R^{-1} \Gamma^T P \Phi - P + Q \\
&= (\Phi - \Gamma K)^T P (\Phi - \Gamma K) - P + Q + K^T \Gamma^T P \Phi + \Phi^T P \Gamma K \\
&\quad - K^T \Gamma^T P \Gamma K - \Phi^T P \Gamma R^{-1} \Gamma^T P \Phi \\
&= (\Phi - \Gamma K)^T P (\Phi - \Gamma K) - P + Q - (RK - \Gamma^T P \Phi)^T R^{-1} (RK - \Gamma^T P \Phi) \\
&\leq (\Phi - \Gamma K)^T P (\Phi - \Gamma K) - P + Q < 0,
\end{aligned} \tag{7.9}$$

where $R = \Gamma^T P \Gamma$.

Now, define a transformation matrix, $T_r := \left[I_{(n-m)} \; -P_{12} P_{22}^{-1}\right]$, and, pre- and post multiply (7.9) by T_r and T_r^T, respectively. Then, this leads to

$$\begin{aligned} 0 &> T_r\left(\Phi^T P\Phi - \Phi^T P\Gamma R^{-1}\Gamma^T P\Phi - P + Q\right) T_r \\ &= (\Phi_{11} - \Phi_{12}P_{22}^{-1}P_{12}^T)^T P_r (\Phi_{11} - \Phi_{12}P_{22}^{-1}P_{12}^T) - P_r + Q_r, \end{aligned} \tag{7.10}$$

where $P_r = P_{11} - P_{12}P_{22}^{-1}P_{12}^T$, and $Q_r = T_r Q T_r^T$. This explicitly shows that the matrix $P_{22}^{-1}P_{12}^T$ is the stabilizing state feedback matrix for the reduced order system by Theorem 7.2.1 since the matrix P_r is positive-definite for the matrix $Q_r \geq 0$.

(2nd argument): The stable sliding coefficient matrix $c_1 \in \mathbb{R}^{m\times(n-m)}$ guarantees the existence of a matrix $P_r > 0$ satisfying

$$(\Phi_{11} - \Phi_{12}c_1)^T P_r (\Phi_{11} - \Phi_{12}c_1) - P_r + \overline{Q}_{11} + \delta I_{n-m} < 0 \tag{7.11}$$

for a scalar $\delta > 0$ and the matrix $\overline{Q}_{11} \geq 0$, where $\overline{Q} := TQT^T$,

$$\overline{Q} := \begin{bmatrix} \overline{Q}_{11} & \overline{Q}_{12} \\ \overline{Q}_{12}^T & \overline{Q}_{22} \end{bmatrix} \in \begin{bmatrix} \mathbb{R}^{(n-m)\times(n-m)} & \mathbb{R}^{(n-m)\times m} \\ \mathbb{R}^{m\times(n-m)} & \mathbb{R}^{m\times m} \end{bmatrix}, \tag{7.12}$$

and

$$T := \begin{bmatrix} I_{n-m} & -c_1^T \\ 0_{m\times(n-m)} & I_m \end{bmatrix}. \tag{7.13}$$

With the matrices P_r and c_1 from (7.11), let a matrix $0 < P := \begin{bmatrix} P_{11} & P_{12} \\ P_{12}^T & P_{22} \end{bmatrix}$ be chosen with the components:

$$\begin{aligned} P_{22} &= \varepsilon I_m + \overline{Q}_{22} + \Phi_{12}^T P_r \Phi_{12} + \tfrac{1}{\delta}(\Phi_r^T P_r \Phi_{12} + \overline{Q}_{12})^T (\Phi_r^T P_r \Phi_{12} + \overline{Q}_{12}), \\ P_{11} &= P_r + c_1^T P_{22} c_1, \\ P_{12} &= c_1^T P_{22} \end{aligned} \tag{7.14}$$

for an $\varepsilon > 0$, where $\Phi_r := \Phi_{11} - \Phi_{12}c_1$. It can be seen that $P > 0$ because $P_{22} > 0$ and

$$\begin{aligned} M &:= P_{11} - P_{12}P_{22}^{-1}P_{12}^T \\ &= P_r + c_1^T P_{22} c_1 - c_1^T P_{22} P_{22}^{-1} P_{22} c_1 \\ &= P_r > 0, \end{aligned}$$

by Lemma 1 introduced in Chapter 6.

Now, let the matrix $K := [K_1 \ K_2]$ be defined with

$$K_1 = \Gamma_2^{-1}(\Phi_{21} + c_1\Phi_{11}) \tag{7.15}$$

and

$$K_2 = \Gamma_2^{-1}(\Phi_{22} + c_1\Phi_{12}). \tag{7.16}$$

Using the equations from (7.11) to (7.16), it holds that

$$\begin{aligned} & T\left[(\Phi - \Gamma K)^T P(\Phi - \Gamma K) - P + Q\right] T^T \\ &= \begin{bmatrix} \Phi_r^T P_r \Phi_r - P_r + \overline{Q}_{11} & \Phi_r^T P_r \Phi_{12} + \overline{Q}_{12} \\ \Phi_{12}^T P_r \Phi_r + \overline{Q}_{12}^T & \Phi_{12}^T P_r \Phi_{12} + \overline{Q}_{22} - P_{22} \end{bmatrix} \end{aligned}$$

$$\leq \begin{bmatrix} -\delta I_{n-m} & \Phi_r^T P_r \Phi_{12} + \overline{Q}_{12} \\ \Phi_{12}^T P_r \Phi_r + \overline{Q}_{12}^T & \Phi_{12}^T P_r \Phi_{12} + \overline{Q}_{22} - P_{22} \end{bmatrix}$$
$$:= \begin{bmatrix} M_{11} & M_{12} \\ M_{12}^T & M_{22} \end{bmatrix}. \tag{7.17}$$

Observe that $M < 0$ since $M_{11} < 0$ and

$$M_{22} - M_{12}^T M_{11}^{-1} M_{12} = -\varepsilon I_m < 0. \tag{7.18}$$

Hence, given a matrix $Q \geq 0$, the matrices $P > 0$ and K lead to the inequality

$$(\Phi - \Gamma K)^T P(\Phi - \Gamma K) - P + Q < 0, \tag{7.19}$$

which proves the second argument. □

Theorem 7.2.2 explicitly provides a systematic means to parameterize all the feasible sliding hyperplanes with the full order Lyapunov matrices. Moreover, the result allows a convenience in practice because the full order original system can be directly used instead of subsystems and the LMIs tool can be simply adopted for computation. Also, from a theoretical point of view, the stabilizability of the full order system is the necessary and sufficient condition for the existence of discrete-time sliding hyperplanes, which has not been clearly pointed out in literature.

Now, based on the results of Theorem 7.2.2, the performance optimization on the sliding hyperplane (i.e., during sliding mode) can be addressed. To this end, the system is assumed to be free of disturbance: $w(k) = 0,\ \forall\, k \geq 0$. Consider the cost index

$$J := \sum_{k=k_s}^{\infty} x(k)^T Q x(k), \tag{7.20}$$

where k_s is the time index to start the sliding mode, i.e., $s(k) = 0,\ \forall\, k \geq k_s$. On the sliding hyperplane, it can be shown that the cost index can be upper bounded by the following theorem.

Theorem 7.2.3. *Suppose that, given a $Q \geq 0$, a sliding function coefficient is chosen by* (7.8) *for $P \in \Omega(Q)$. Then, it holds that*

$$J := \sum_{k=k_s}^{\infty} x(k)^T Q x(k) < \overline{\sigma}(P) \cdot \| x(k_s) \|^2 \tag{7.21}$$

Proof. For K satisfying (7.7), let the system equation be rewritten as follows:

$$x(k+1) = \Phi_c x(k) + \Gamma \delta(k), \tag{7.22}$$

where $\Phi_c = \Phi - \Gamma K,\ \delta(k) = u(k) + Kx(k)$. Then, for a quadratic function, $V(k) = x(k)^T P x(k)$, it can be shown that

$$V(k+1) - V(k) = x(k)^T \left(\Phi_c^T P \Phi_c - P\right) x(k) + 2x(k)^T \Phi_c^T P \Gamma \delta(k) + \delta(k)^T \Gamma^T P \Gamma \delta(k).$$

Considering that, from (7.22),

$$\Phi_c x(k) = x(k+1) - \Gamma\delta(k),$$

it leads to

$$V(k+1)-V(k)=x(k)^T\left(\Phi_c^T P\Phi_c - P\right)x(k)+2x(k+1)^T P\Gamma\delta(k)-\delta(k)^T\Gamma^T P\Gamma\delta(k).$$

Now, from (7.4), note that, with $c_1 = P_{22}^{-1}P_{12}^T$ and $\Gamma^T = [0,\ \Gamma_2^T]$,

$$\begin{aligned} s(k) &= x_2(k) + c_1 x_1(k) \\ &= P_{22}^{-1}(P_{12}^T x_1(k) + P_{22}x_2(k)) \\ &= (\Gamma_2 P_{22})^{-T}\Gamma^T P x(k), \end{aligned} \tag{7.23}$$

which implies that $\Gamma^T Px(k) = 0,\ \forall\, k \geq k_s$ for the instance k_s in which the sliding mode starts. Therefore, it follows that, for $k \geq k_s$,

$$\begin{aligned} V(k+1) - V(k) &= x(k)^T\left(\Phi_c^T P\Phi_c - P\right)x(k) - \delta(k)^T\Gamma^T P\Gamma\delta(k) \\ &< -x(k)^T Qx(k). \end{aligned} \tag{7.24}$$

By taking the summation of both sides, it can be seen that

$$\sum_{k=k_s}^{\infty} x(k)^T Qx(k) < \overline{\sigma}(P)\cdot \parallel x(k_s) \parallel^2 . \tag{7.25}$$

This completes the proof. □

Therefore, the design steps for the discrete-time sliding hyperplane can be summarized as follows.

Sliding Hyperplane Design Procedure

1. Select a matrix $Q \geq 0$ considering the relative state weight.
2. Solve the minimization problem to have P^*:

$$P^* = \min_{P\in\Omega(Q)} \overline{\sigma}(P) \tag{7.26}$$

3. Compute the sliding function coefficient such that $c_1 = P_{22}^{*-1}P_{12}^{*T}$.

It is noted that the discrete-time Lyapunov inequality in (7.7) can be readily converted into the LMIs and solved by numerical tools, e.g., LMIs Toolbox of MATLAB.

7.3 Reaching Law Design

To accomplish the sliding mode, the system state, which may not be on the sliding hyperplane, should be forced to reach the sliding hyperplane. However, it has been known that the perfect sliding mode (i.e., $s(k+1)=s(k)=0$) cannot be achieved in DSMC since the advantage of the infinite switching capability cannot be taken. To address this problem, the concept of a quasi-sliding mode has been effectively adopted in literature.

To derive the control law that drives the sliding function within a bound, consider the control law

$$u(k)=-(c^T\Gamma)^{-1}\{c^T\Phi x(k)-\widetilde{\beta}\cdot s(k)+\widetilde{z}\cdot f(s(k),\phi)\}, \quad (7.27)$$

where $c^T := [c_1,\ I_m] \in \mathbb{R}^{m\times n}$, $\widetilde{\beta}=diag[\beta_1,\cdots,\beta_m]$, $\widetilde{z}=diag[z_1,\cdots,z_m]$, $\phi=[\phi_1,\cdots,\phi_m]$ and

$$f(s(k),\phi) := \left[sat\left(\frac{s_1(k)}{\phi_1}\right),\cdots,sat\left(\frac{s_m(k)}{\phi_m}\right)\right]^T \in \mathbb{R}^m$$

for scalar parameters z_i and $\phi_i>0$ satisfying

$$z_i=\sum_{j=1}^{l}|(c^TF)_{ij}|\overline{w}_j \quad (7.28)$$

and

$$\frac{2z_i}{\phi_i}-1<\beta_i<1, \quad (7.29)$$

which necessarily implies that $\phi_i>z_i$.

With the control law (7.27) and (7.4), it can be seen that

$$\begin{aligned} s(k+1) &= c^Tx(k+1)\\ &= c^T\Phi x(k)+c^T\Gamma u(k)+c^TFw(k)\\ &= -\widetilde{\beta}s(k)-\widetilde{z}\cdot f(s(k),\phi)+c^TFw(k), \end{aligned} \quad (7.30)$$

which results in

$$s_i(k+1)=\beta_i s_i(k)-z_i sat\left(\frac{s_i(k)}{\phi_i}\right)+d_i(k), \quad (7.31)$$

where

$$d_i(k)=\sum_{j=1}^{l}(c^TF)_{ij}w_j(k).$$

To show the quasi-sliding mode behavior, let us consider the Lyapunov function, $V_i(k)=s_i^2(k)$. Then, it follows that

$$\begin{aligned} \Delta V_i(k) &= s_i^2(k+1)-s_i^2(k)\\ &= \Delta s_i(k)\left(\Delta s_i(k)+2s_i(k)\right), \end{aligned}$$

where $\Delta s_i(k)=s_i(k+1)-s_i(k)$.

First, consider the case when $s_i(k) > \phi_i$. From (7.31), it holds that

$$\begin{aligned}\Delta s_i(k) &= (\beta_i - 1)s_i(k) - z_i + d_i(k) \\ &< (\beta_i - 1)\phi_i - z_i + \Sigma_{j=1}^{l} |(c^T F)_{ij}|\overline{w}_j \\ &= (\beta_i - 1)\phi_i < 0\end{aligned} \tag{7.32}$$

since $\beta_i - 1 < 0$. Moreover,

$$\begin{aligned}(\Delta s_i(k) + 2s_i(k)) &= (\beta_i + 1)s_i(k) - z_i + d_i(k) \\ &\geq (\beta_i + 1)s_i(k) - 2z_i \\ &> \tfrac{2z_i}{\phi_i} \times \phi_i - 2z_i = 0,\end{aligned}$$

which shows that $\Delta V_i(k) < 0$. In a similar way, it can be shown that $\Delta V_i(k) < 0$ in the case of $s_i(k) < -\phi_i$.

Now, suppose that $|s_i(k)| < \phi_i$. From (7.31), it holds that

$$s_i(k+1) = \left(\beta_i - \frac{z_i}{\phi_i}\right) s_i(k) + d_i(k). \tag{7.33}$$

Using the fact that, from the condition (7.29),

$$\left|\beta_i - \frac{z_i}{\phi_i}\right| < \frac{\phi_i - z_i}{\phi_i}, \tag{7.34}$$

we can immediately obtain

$$\begin{aligned}|s_i(k+1)| &< \left|\beta_i - \tfrac{z_i}{\phi_i}\right| |s_i(k)| + |d_i(k)| \\ &< \tfrac{\phi_i - z_i}{\phi_i} \times \phi_i + z_i = \phi_i.\end{aligned}$$

This implies that the sliding function should remain inside the bound once it enters the bound.

Overall, compared to the results in Chapter 6, it can be seen that the results obtained for DSMC are similar to those for the continuous-time SMC. As to the sliding hyperplane design, the Lyapunov inequalities for the nominal full order system can be used to parameterize all the stable sliding hyperplanes either in the continuous-time case or in the discrete-time case. As to the reaching law design, the saturation function is used instead of the signum function in order to relax the convergence bound of the sliding function in the case of DSMC.

7.4 DSMC with Multi-Rate Output Feedback

7.4.1 State Estimate via Multi-Rate Output

In section 3.2, multi-rate output feedback is reviewed for deterministic system. In this section, MROF is reviewed for an uncertain system [8]. Consider an uncertain system in which only output information is available.

$$\begin{cases} x(k+1) = \Phi x(k) + \Gamma u(k) + F w(k) \\ y(k) = C x(k), \end{cases} \tag{7.35}$$

where $y \in \mathbb{R}^h$ is the measurement vector. Note that (7.35) is obtained by sampling the continuous-time system at every τ seconds. In order to apply the multi-rate output feedback approach introduced in [53], it is assumed that high rate sampling with the time interval $\Delta = \tau/N$ for an integer N is available, where N is chosen to be equal to or larger than the observability index of the pair (Φ, C). Then, the state space description for the sub-sampled system (i.e., Δ-system) is given by

$$x(q+1) = \Phi_\Delta x(q) + \Gamma_\Delta u(q) + F_\Delta w(q), \tag{7.36}$$

where

$$\Phi = \Phi_\Delta^N,\ \Gamma = \sum_{i=0}^{N-1} \Phi_\Delta^i \Gamma_\Delta,\ F = \sum_{i=0}^{N-1} \Phi_\Delta^i F_\Delta.$$

Thus, the multi-rate output vector can be expressed as

$$y_k = C_0 x(k) + D_0 u(k) + C_d w(k), \tag{7.37}$$

where

$$C_0 = \begin{bmatrix} C \\ C\Phi_\Delta \\ C\Phi_\Delta^2 \\ \vdots \\ C\Phi_\Delta^{N-1} \end{bmatrix},\ D_0 = \begin{bmatrix} 0 \\ C\Gamma_\Delta \\ C\Phi_\Delta\Gamma_\Delta + C\Gamma_\Delta \\ \vdots \\ C\sum_{i=0}^{N-2}\Phi_\Delta^i\Gamma_\Delta \end{bmatrix},\ C_d = \begin{bmatrix} 0 \\ CF_\Delta \\ C\Phi_\Delta F_\Delta + CF_\Delta \\ \vdots \\ C\sum_{i=0}^{N-2}\Phi_\Delta^i F_\Delta \end{bmatrix} \tag{7.38}$$

and $y_k = \left[y((k-1)T)\ y((k-1)T+\Delta)\ \cdots\ y(kT-\Delta) \right]^T$. From (7.35) and (7.37), $x(k)$ can be obtained from the output samples as follows:

$$x(k) = L_y y_k + L_u u(k-1) + L_d w(k-1), \tag{7.39}$$

where

$$\begin{aligned} L_y &= \Phi \left(C_0^T C_0\right)^{-1} C_0^T, \\ L_u &= \Gamma - L_y D_0, \\ L_d &= F - L_y C_d. \end{aligned} \tag{7.40}$$

Note that (7.39) provides a static state estimate from the multi-rate output samples y_k as follows:

$$\bar{x}(k) = L_y y_k + L_u u(k-1). \tag{7.41}$$

Remark 2. As a matter of fact, a more accurate state estimate can be obtained by incorporating the mean value of the disturbance as follows:

$$\bar{x}(k) = L_y y_k + L_u u(k-1) + l_0, \tag{7.42}$$

where

$$l_{l,i} < (L_d w(k-1))_i < l_{u,i},\ l_0 = \frac{l_l + l_u}{2},$$

when the upper and the lower bounds are available. However, for simplicity of description, assume that l_0 is equal to (or very near to) zero since the known disturbance bias can be easily canceled out by static compensation.

7.4.2 Multi-Rate Output Feedback Reaching Law Design

With the state estimate (7.41), let the sliding function *measurable* with partial measurements be defined by

$$\bar{s}(k) = c^T \bar{x}(k) = c^T L_y y_k + c^T L_u u(k-1). \tag{7.43}$$

To achieve the quasi-sliding mode, consider the reaching law obtained from (7.27) by simple replacements of $x(k)$ and $s(k)$ by $\bar{x}(k)$ and $\bar{s}(k)$, respectively, as

$$u(k) = -(c^T \Gamma)^{-1}\{c^T \Phi \bar{x}(k) - \widetilde{\beta} \cdot \bar{s}(k) + \widetilde{z} \cdot f(\bar{s}(k), \phi)\}, \tag{7.44}$$

where (7.29) holds with

$$z_i = \sum_{j=1}^{l} \{|(c^T \Phi L_d)_{ij}| + |(c^T L_y C_d)_{ij}|\} \overline{w}_j. \tag{7.45}$$

As in the previous section, for the Lyapunov function $V_i(k) = \bar{s}_i^2(k)$, the ultimate target is to show the stability of $\bar{s}(k)$ by proving the decrement of the Lyapunov function:

$$V_i(k+1) - V_i(k) = [2\bar{s}_i(k) + \Delta\bar{s}_i(k)]\,\Delta\bar{s}_i(k). \tag{7.46}$$

Using the relationship such that, from (7.41) and (7.39),

$$\bar{x}(k) = x(k) - L_d w(k-1),$$

the incremental change in the sliding function can be obtained as

$$\begin{aligned} \bar{s}(k+1) - \bar{s}(k) &= c^T (x(k+1) - L_d w(k)) - \bar{s}(k) \\ &= c^T [\Phi x(k) + \Gamma u(k) + F w(k) - (F - L_y C_d) w(k)] - \bar{s}(k) \\ &= (\widetilde{\beta} - I)\bar{s}(k) + d(k) - \widetilde{z} \cdot f(\bar{s}(k), \phi), \end{aligned} \tag{7.47}$$

where

$$d(k) = c^T \Phi L_d w(k-1) + c^T L_y C_d w(k).$$

Hence, this leads to

$$\Delta\bar{s}_i(k) = (\beta_i - 1)\bar{s}_i(k) + d_i(k) - z_i \mathrm{sat}\left(\frac{\bar{s}_i(k)}{\phi_i}\right), \tag{7.48}$$

where

$$d_i(k) = \sum_{j=1}^{l} \left\{ (c^T \Phi L_d)_{ij} w_j(k-1) + (c^T L_y C_d)_{ij} w_j(k) \right\}.$$

Now, consider the case of $\bar{s}_i(k) > \phi_i$. Using the fact that $d_i(k) \le z_i$, it can be shown that

$$\begin{aligned} \Delta \bar{s}_i(k) &\le (\beta_i - 1)\bar{s}_i(k) + z_i - z_i \\ &= (\beta_i - 1))\bar{s}_i(k) < 0 \end{aligned} \tag{7.49}$$

since $\beta_i - 1 < 0$. Then, observe that

$$\begin{aligned} [2\bar{s}_i(k) + \Delta \bar{s}_i(k)] &= (\beta_i + 1)\bar{s}_i(k) + d_i(k) - z_i \\ &\ge (\beta_i + 1)\bar{s}_i(k) - 2z_i \\ &> (\beta_i + 1)\phi_i - 2z_i \ge 0 \end{aligned} \tag{7.50}$$

thanks to the constraint (7.29). So, $\bar{s}_i(k)$ decreases as k increases.

In the case of $\bar{s}_i(k) < -\phi$, the proof procedures are very similar to those above. Thus, they are omitted here in order to save space.

When $|\bar{s}_i(k)| < \phi_i$, it may be shown from (7.48) that

$$\bar{s}_i(k+1) = \left(\beta_i - \frac{z_i}{\phi_i} \right) \bar{s}_i(k) + d_i(k). \tag{7.51}$$

Thus, with (7.29), it holds that

$$\begin{aligned} |\bar{s}_i(k+1)| &\le \left| \beta_i - \tfrac{z_i}{\phi_i} \right| \cdot |\bar{s}_i(k)| + z_i \\ &< \tfrac{\phi_i - z_i}{\phi_i} \cdot \phi_i + z_i = \phi_i, \end{aligned} \tag{7.52}$$

which proves that $\bar{s}(k)$ remains within the band ϕ once the trajectory enters the band.

Now, to investigate the *true* sliding function behavior, observe that

$$\bar{s}(k) - s(k) = c^T (\bar{x}(k) - x(k)) = -c^T L_d w(k-1), \tag{7.53}$$

which yields

$$|s_i(k)| < \phi_i + \psi_i, \tag{7.54}$$

where

$$\psi_i = \sum_{j=1}^{l} |(c^T L_d)_{ij}| \bar{w}_j.$$

Considering that ϕ_i should be chosen in relation to the disturbance magnitude in (7.29), the above result implies that the sliding function $s(k)$ remains within an expanded band depending on the magnitude of external disturbances.

7.5 Numerical Simulation

Let the motion stabilization problem in Fig. 7.1 be considered, which is adopted from [64]. Assuming that $m_1 = m_2 = m_3 = 1(kg)$, $k_1 = k_2 = 3.3(N/m)$, the continuous-time system is given by

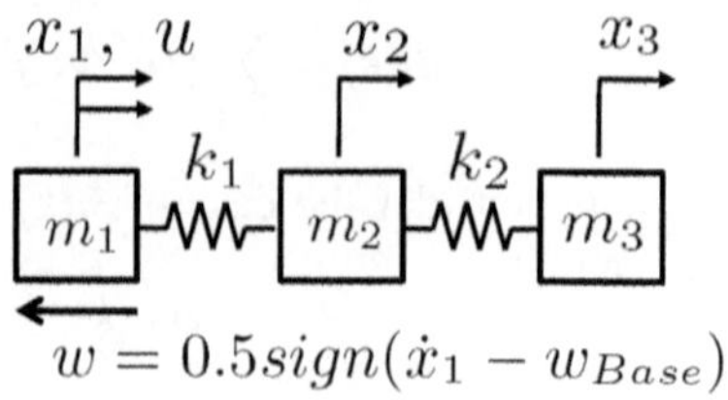

Fig. 7.1 An example of the motion stabilization problem. $w_{Base} = sin(2\pi t)$.

$$\dot{x}_c = \left[\begin{array}{ccc|c} & 0_{3\times3} & & I_{3\times3} \\ \hline -3.3 & 3.3 & 0 & \\ 3.3 & -6.6 & 3.3 & 0_{3\times3} \\ 0 & 3.3 & -3.3 & \end{array}\right] x_c + \left[\begin{array}{c} 0_{3\times1} \\ \hline 1 \\ 0 \\ 0 \end{array}\right] u + \left[\begin{array}{c} 0_{3\times1} \\ \hline -1 \\ 0 \\ 0 \end{array}\right] w$$

$$y = \begin{bmatrix} 0 & 0 & 1 & 0 & 0 & 0 \end{bmatrix} x_c,$$

where $x_c := [x_1,\ x_2,\ x_3,\ \dot{x}_1,\ \dot{x}_2,\ \dot{x}_3]^T$. Without the feedback stabilization control, the frictional disturbance, w, between the contact surfaces of mass 1 and the base would cause the vibration of the masses, as shown in Fig. 7.2.

With $\tau = 0.06(sec)$, to obtain the sampled data system in the regular form of (7.35), the MATLAB function –`c2dm(·)`– was used ,and, then, the state coordinate transformation, $x(k) = T_r x_c(k)$, was performed to achieve the regular form system as follows:

$$\Phi = \begin{bmatrix} 9.88\times10^{-1} & 5.92\times10^{-3} & -6.04\times10^{-3} & 5.98\times10^{-2} & 1.19\times10^{-4} & 4.13\times10^{-4} \\ 5.92\times10^{-3} & 9.94\times10^{-1} & -6.34\times10^{-6} & 1.19\times10^{-4} & 5.99\times10^{-2} & 7.27\times10^{-7} \\ 7.47\times10^{-4} & -3.72\times10^{-4} & 9.99\times10^{-1} & 1.96\times10^{-4} & -1.12\times10^{-5} & -6.00\times10^{-2} \\ -3.94\times10^{-1} & 1.97\times10^{-1} & -1.97\times10^{-1} & 9.88\times10^{-1} & 5.92\times10^{-3} & 1.17\times10^{-2} \\ 1.97\times10^{-1} & -1.97\times10^{-1} & -4.08\times10^{-4} & 5.92\times10^{-3} & 9.94\times10^{-1} & 3.51\times10^{-5} \\ 1.96\times10^{-1} & 7.81\times10^{-4} & 1.97\times10^{-1} & 6.30\times10^{-3} & 1.78\times10^{-5} & 9.90\times10^{-1} \end{bmatrix},$$

$$\Gamma = \begin{bmatrix} 0 \\ 0 \\ 0 \\ 0 \\ 0 \\ 0.0599 \end{bmatrix},\ F = \begin{bmatrix} 0 \\ 0 \\ 0 \\ 0 \\ 0 \\ -0.0599 \end{bmatrix},$$

$$C = \begin{bmatrix} -3.39\times10^{-13} & 1 & -1.14\times10^{-8} & -2.26\times10^{-11} & -1.34\times10^{-14}, & 1.18\times10^{-8} \end{bmatrix}.$$

To define the quadratic cost index in (7.20), let the weighting matrix be defined as $Q = T_r^{-T} Q_x T_r^{-1}$, which yields

$$J = \sum_{k=k_s}^{\infty} x(k)^T Q x(k) = \sum_{k=k_s}^{\infty} x_c(k)^T Q_x x_c(k), \tag{7.55}$$

for imposing the relative weights to each of physical state in x_c. For the sliding hyperplane design, the weight matrix is chosen as

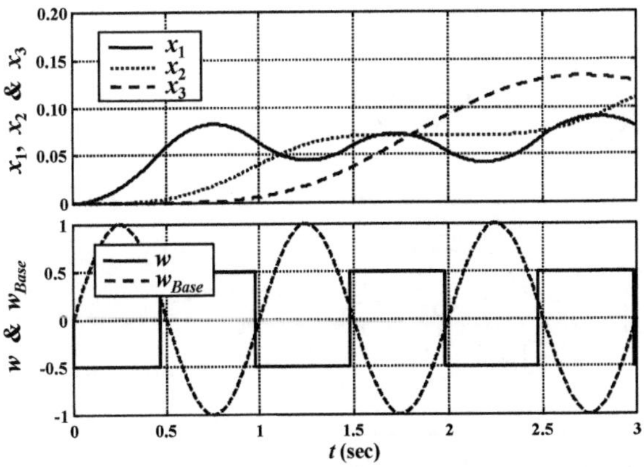

Fig. 7.2 Destabilization of masses without control.

$$Q_x = diag[0,\ 0.5,\ 1,\ 0,\ 0.3,\ 0.5] \times 10^{-6}.$$

Using the LMIs solver in MATLAB, the minimization of $\overline{\sigma}(P)$ under the Lyapunov inequality in Theorem 7.2.2 could be easily solved, and, the optimal Lyapunov matrix P^* was used for generating the sliding function coefficient

$$c_1 = [19.0,\ -9.07,\ -6.38,\ 12.8,\ 9.81].$$

For the full state feedback design, the selected parameters are as follows:

$$\phi = 0.0330,\ \beta = 0.909,\ z = 0.0300.$$

The simulation results are shown in Fig. 7.3. The positions of the masses are suitably stabilized compared to the no-feedback case shown in Fig. 7.2. Also, note that the upper bound analysis for the sliding function in (7.35) is not very conservative in this example as can be seen in Fig. 7.3.

To see the multi-rate output feedback performance, the multi-rate simulation was carried out with the number of sub-samples $N = 6$ (i.e., $\Delta = 0.01(sec)$). Following the design rules in (7.44) and (7.45), the parameters are chosen as follows:

$$\phi = 0.0425,\ \beta = 0.909,\ z = 0.0386.$$

Also, from (7.54), it can be shown that $\psi = 0.0157$, which gives a upper bound of sliding function magnitude of $\phi + \psi = 0.0582$. As shown in Fig. 7.4, the regulation performance slightly degrades compared to the full state feedback case. Also, the upper bound of the actual sliding function could be estimated by (7.54) non-conservatively.

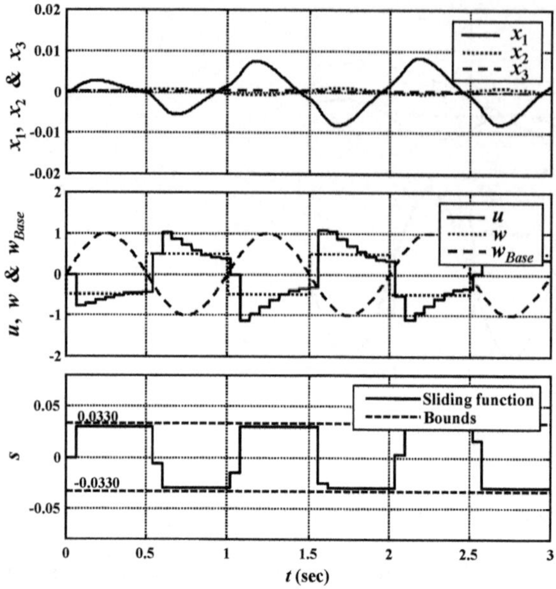

Fig. 7.3 Vibration suppression using full state measurements.

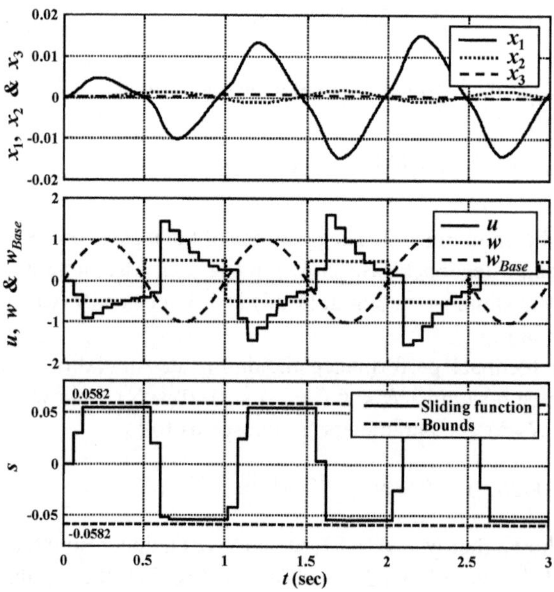

Fig. 7.4 Vibration suppression using multi-rate output feedback.

7.6 Conclusion

In this chapter, a complete sequence to design the discrete-time sliding mode control have been proposed for a class of uncertain LTI discrete-time systems. For the sliding hyperplane design, it was shown that the discrete-time Lyapunov inequality for the original full order system can be utilized for parameterizing all the stable sliding hyperplanes. This also allows us to define an optimization of a guaranteed cost on the sliding mode. By adopting the quasi-sliding mode approach, the reaching law was derived, which admits the sliding function within a certain band. In order to relax the full state requirement, it was demonstrated that the multi-rate output feedback can be successfully adopted while keeping the sliding function within an expanded bound. Through the simulations, the validity and effectiveness of the proposed scheme was verified.

References

1. Bandyopadhyay, B., Fulwani, D.: Tracking control for discrete SISO system with non-linear sliding surface. In: 3rd National Conference on Instrumentation and Control Engineering ICECON 2007, Trichy, India, pp. 63–67 (2007)
2. Bandyopadhyay, B., Fulwani, D.: A robust tracking controller for uncertain MIMO plant using nonlinear sliding surface. In: Proceedings of IEEE International Conference on Industrial Technology (ICIT 2009), Gippsland, Australia, pp. 1–6 (2009)
3. Bandyopadhyay, B., Fulwani, D.: High performance tracking controller for discrete plant using non-linear sliding surface. IEEE Transaction on Ind. Electronics (to appear, 2009)
4. Bandyopadhyay, B., Fulwani, D., Park, Y.J.: A robust CNF controller using integral sliding mode with application to a servo system. Under review in Control Engineering and Practice
5. Bandyopadhyay, B., Fulwani, D., Park, Y.J.: A robust algorithm against actuator saturation using integral sliding mode and composite nonlinear feedback. In: Proceedings of the 17th IFAC World Congress, Seoul, Korea, pp. 14,174–14,179 (2008)
6. Bandyopadhyay, B., Fulwani, D., Park, Y.J.: A robust algorithm against actuator saturation using integral sliding mode and composite nonlinear feedback. In: Proc. 17th IFAC World Congress, Seoul, Korea, pp. 14174–14179 (2008)
7. Bandyopadhyay, B., Fulwani, D., Postlethwaite, I., Turner, M.: A nonlinear sliding surface to improve performance of discrete-time input-delay system. Under review in International Journal of Control
8. Bandyopadhyay, B., Janardhanan, S.: Discrete-Time Sliding Mode Control. LNCIS, vol. 323. Springer, Heidelberg (2006)
9. Barnett, S.: Polynomials and Linear Control Systems. Marcel Dekker Inc., New York (1983)
10. Bartolini, G., Ferrara, A., Utkin, V.I.: Adaptive sliding mode control in discrete-time systems. Automatica 31(5), 679–773 (1995)
11. Bartoszewicz, A.: Time-varying sliding modes for second-order systems. IEE Proc.-Control Theory Appl. 143(5), 455–462 (1996)
12. Bartoszewicz, A.: Discrete-time quasi-sliding-mode control strategies. IEEE Trans. on Industrial Electronics 45(3), 633–637 (1998)
13. Bartoszewicz, A., Kaynak, O., Utkin, V.I.: (eds.) special section on sliding mode control in industrial applications. IEEE Transactions on Industrial Electronics 55(11), 3806–4074 (2008)

14. Bartoszewicz, A., Nowacka-Leverton, A.: Time-Varying Sliding Modes for Second and Third Order Systems. LNCIS, vol. 382, pp. 1–200. Springer, Heidelberg (2009)
15. Bartoszewicz, A., Patton, R.J. (eds.) Sliding mode control: Special issue. Int. Journal of Adaptive Control and Signal Processing 21, 635–822 (2007)
16. Bejarano, F.J., Fridman, L., Poznyak, A.: Output integral sliding mode control based on algebraic hierarchical observer. Int. J. of Control 80(3), 443–453 (2006)
17. Boyd, S., Ghaoui, L.E., Feron, E., Balarkrishnan, V.: Linear Matrix Inequalities in Systems and Control Theory. SIAM Studies in Applied Mathematics 15 (1994)
18. Carpita, M., Marchesoni, M.: Experimental study of a power conditioning system using sliding mode control. IEEE Transactions on Power Electronics 11(5), 731–742 (1996)
19. Castanos, F., Fridman, L.: Analysis and design of integral sliding manifolds for systems with unmatched perturbations. IEEE Trans. on Automatic Control 51(5), 853–858 (2006)
20. Chen, B.M., Cheng, G., Lee, T.H.: Modeling and compensation of nonlinearities and friction in a micro hard disk servo system with nonlinear control. IEEE Trans. on Control Systems Technology 13(5), 708–721 (2005)
21. Chen, B.M., Lee, T.H., Peng, K., Venkataramanan, V.: Composite nonlinear feedback control for linear systems with input saturation: theory and an applications. IEEE Trans. on Automatic Control 48(3), 427–439 (2003)
22. Cheng, G., Peng, K.: Robust composite nonlinear feedback control with application to a servo positioning system. IEEE Transaction on Ind. Electronics 54(2), 1132–1140 (2007)
23. Chiasson, J.: Modeling and High Performance Control of Electric Machinery. Marcel Dekker, New York (1998)
24. Chilali, M., Gahinet, P.: H_∞ design with pole placement constraints: an LMI approach. IEEE Trans. on Automatic Control 41, 358–367 (1996)
25. Choi, H.-H.: An explicit formula of linear sliding surfaces for a class of uncertain dynamic systems with mismatched uncertainties. Automatica 34, 1015–1020 (1998)
26. Choi, S.-B., Cheong, C.C., Park, D.-W.: Moving switching surfaces for robust control of second-order variable structure systems. International Journal of Control 58, 229–245 (1993)
27. Choi, S.-B., Park, D.-W., Jayasuriya, S.: A time-varying sliding surface for fast and robust tracking control of second order uncertain systems. Automatica 30(5), 899–904 (1994)
28. Corradini, M.L., Orlando, G.: Linear unstable plants with saturating actuators: Robust stabilization by a time varying sliding surface. Automatica 43, 88–94 (2007)
29. Cortés, J.: Discontinuous dynamical system- a tutorial on solutions, nonsmooth analysis, and stability. IEEE Control System Magazine June, 36–73 (2008)
30. DeCarlo, R.A., Zak, S.H., Matthews, G.P.: Variable structure control of nonlinear multivariable systems: a tutorial. Proceedings of the IEEE 76, 212–232 (1988)
31. Dorling, C.M., Zinober, A.S.I.: Two approaches to hyperplane design in multivariable structure control systems. Int. Journal of Control 44, 65–82 (1986)
32. Educational Control Products, California, USA: Installation and user manual of ECP model 220 for use with MATLAB- R14 using Real-Time Windows Target (RTWT), version 3.0 (2004)
33. Edwards, C., Akoachere, A., Spurgeon, S.K.: Sliding-mode output feedback controller design using linear matrix inequalities. IEEE Transactions on Automatic Control 46(1), 115–119 (2001)
34. Edwards, C., Spurgeon, S.K.: Sliding mode stabilization of uncertain systems using only output information. International Journal of Control 62(5), 1129–1144 (1995)

35. Edwards, C., Spurgeon, S.K.: Robust output tracking using a sliding-mode controller/observer scheme. International Journal of Control 64(5), 967–983 (1996)
36. Edwards, C., Spurgeon, S.K.: Sliding Mode Control: Theory and Application. Taylor and Francis, UK (1998)
37. EL-Ghezawi, O.M.E., Zinober, A.S.I., Billings, S.A.: Analysis and design of variable structure systems using a geometric approach. International Journal of Control 38(3), 657–671 (1983)
38. Emel'yanov, S.V.: Method of designing complex control algorithm using an error and its first time derivative only. Automation and Remote Control 18(10) (1957) (in Russian)
39. Emel'yanov, S.V., Burovoi, I.A., et al.: Mathematical models of process in technology and development of variable structure control system. Metallurgy (Moscow) 18(07) (1964) (in Russian)
40. Eun, Y., Kim, J.H., Cho, D.I.: Discrete-time variable structure controller with a decoupled disturbance compensator and its application to a CNC servomechanism. IEEE Trans. on Control Systems Technology 7(4), 414–423 (1999)
41. Filippov, A.F.: Differential equations with discontinuous right hand side. American Mathematical Society Translations 42, 199–231 (1964)
42. Franklin, G., Powell, J.D., Workman, M.L.: Digital Control of Dynamic Systems. Addison-Wesley, Reading (1990)
43. Furukawa, T., Shimemura, E.: Predictive control for systems with time delay. Int. J. of Control 37(2), 399–412 (1983)
44. Furuta, K.: Sliding mode control of a discrete system. Systems and Control Letters 14, 145–152 (1990)
45. Furuta, K., Pan, Y.: Variable structure control with sliding sector. Automatica 36, 211–228 (2000)
46. Gao, W., Wang, Y., Homaifa, A.: Discrete-time variable structure control systems. IEEE Transactions on Industrial Electronics 42(2), 117–122 (1995)
47. Golo, G., Milosavljevi, C.: Robust discrete-time chattering free sliding mode control. Systems and Control Letters 41, 19–28 (2000)
48. Goodwin, G.C., Middleton, R.H., Poor, H.V.: High-speed digital signal processing and control. Proceedings of the IEEE 80(2), 240–259 (1992)
49. He, Y., Chen, B.M., Wu, C.: Improving transient performance in tracking control for linear multivariable discrete-time systems with input saturation. Systems and Control Letters 56, 25–33 (2007)
50. Hu, Q., Du, C., Xie, L., Wang, Y.: Discrete-time sliding mode control with time varying surface for hard disc drives. IEEE Transactions on Control System Technology 17(1), 175–183 (2009)
51. Hung, J.Y., Gao, W., Hung, J.C.: Variable structure control: A survey. IEEE Transaction on Ind. Electronics 40(1), 2–22 (1993)
52. Itkis, U.: Control Systems of Variable Structure. Wiley, New York (1976)
53. Janardhanan, S., Bandyopadhyay, B.: Discrete sliding mode control of systems with unmatched uncertainty using multirate output feedback. IEEE Transactions on Automatic Control 51(6), 1030–1035 (2006)
54. Janardhanan, S., Bandyopadhyay, B.: Output feedback sliding-mode control for uncertain systems using fast output sampling technique. IEEE Transactions on Industrial Electronics 53(5), 1677–1682 (2006)
55. Janardhanan, S., Kariwala, V.: Multirate output feedback based LQ optimal discrete-time sliding mode control. IEEE Transactions on Automatic Control 53(1), 367–373 (2008)

56. Kenjo, T.: Stepping Motors and their Microprocessor Controls. Oxford, New York (1984)
57. Kharitonov, V.L., Zhabkob, A.P.: Lyapunov-Krasovskii approach to the robust stability analysis of time-delay systems. Automatica 39, 15–20 (2003)
58. Kim, K.S., Bandyopadhyay, B.: A Lyapunov approach for discrete-time sliding hyperplane design and robust sliding mode control using MROF. Under revision in Automatica
59. Kim, K.-S., Bandyopadhyay, B.: A Lyapunov approach for discrete-time sliding hyperplane design and robust sliding mode control using MROF. In: Proceedings of the 17th IFAC World Congress, Seoul, Korea, pp. 1129–1134 (2008)
60. Kim, K.-S., Jabbari, F.: Using scales in multi-objective approach. IEEE Trans. Automatic Control 45, 973–977 (2000)
61. Kim, K.S., Park, Y.: Using Lyapunov matrices for sliding mode design. In: Proceedings of the 39th Conference on Decision and Control (CDC), Sydney, Australia, pp. 2204–2209 (2000)
62. Kim, K.-S., Park, Y.: Parametric approaches to sliding mode design for linear multivariable systems. International Journal of Control, Automation and Systems 1(1), 11–18 (2003)
63. Kim, K.-S., Park, Y.: Sliding mode design via quadratic performance optimization with pole-clustering constraint. SIAM Journal of Control and Optimization 49(2), 670–684 (2004)
64. Kim, K.-S., Park, Y., Oh, S.-H.: Designing robust sliding hyperplanes for parametric uncertain systems: a Riccati approach. Automatica 36, 1041–1048 (2000)
65. Kolmanovskii, V., Myshkis, A.: Introduction to the Theory and Applications of Functional Differential Equations. Kluwer Academic Publishers, Dordrecht (1999)
66. Korondi, P., Hashimoto, H., Utkin, V.: Direct torsion control of flexible shaft in an observer-based discrete-time sliding mode. IEEE Transactions on Industrial Electronics 45(2), 291–296 (1998)
67. Kose, I.E., Jabbari, F.: Robust control of linear systems with real parametric uncertainty. Automatica 35(4), 679–687 (1999)
68. Lee, D.-S., Youn, M.-J.: Controller design of variable structure systems with nonlinear sliding surface. Electronics Letters 25(25), 1715–1717 (1989)
69. Levant, A.: Robust exact differentiation via sliding mode technique. Automatica 34, 379–384 (1998)
70. Lin, Z., Pachter, M., Banda, S.: Towards improvement of tracking performance-nonlinear feedback for linear systems. International Journal of Control 70(1), 1–11 (1998)
71. Lozano, R., Castillo, P., Garcia, P., Dzul, A.: Robust prediction-based control for unstable delay systems: application to the yaw control of a mini-helicopter. Automatica 40, 603–612 (2004)
72. Manitius, A.Z., Olbrot, A.W.: Finite spectrum assignment problem for systems with delays. IEEE Trans. on Automatic Control 24(4), 541–553 (1979)
73. Middleton, R.H., Goodwin, G.C.: Improved finite word length characteristics in digital control using delta operators. IEEE Transactions on Automatic Control AC-31(11), 1015–1021 (1986)
74. Middleton, R.H., Goodwin, G.C.: Digital Control and Estimation: A Unified Approach. Prentice-Hall, Englewood Cliffs (1990)
75. Milosavljevi, C.: General conditions for the existence of a quasi-sliding mode on the switching hyperplane in discrete variable structure systems. Automation and Remote Control 3, 36–44 (1985)

76. Milosavljevi, C.: In Variable structure systems from principles to implementation. IEE, U.K (2004)
77. Mirkin, L., Raskin, N.: Every stabilizing dead-time controller has an observer-predictor based structure. Automatica 39, 1747–1754 (2003)
78. Misawa, E.A.: Discrete-time sliding mode control: The linear case. Journal of Dynamic Systems, Measurement, and Control 119(2), 819–821 (1997)
79. Park, D.-W., Choi, S.-B.: Moving sliding surfaces for high-order variable structure systems. International Journal of Control 72, 960–970 (1999)
80. Park, M.-H., Kim, K.-S.: Chattering reduction in the position control of induction motor using the sliding mode. IEEE Transactions on Power Electronics 6(3), 317–325 (1991)
81. Paul, A.K., Mishra, J.K., Radke, M.G.: Reduced order sliding mode control for pneumatic actuator. IEEE Transactions on Control Systems Technology 2(3), 271–276 (1994)
82. Reddy, G.D., Bandyopadhyay, B., Tiwari, A.P.: Multirate output feedback based sliding mode spatial control for a large PHWR. IEEE Trans. on Nuclear Science 54(6), 2677–2686 (2007)
83. Richard, J.-P.: Time-delay systems an overview of some recent advances and open problems. Automatica 39, 1667–1694 (2003)
84. Scherer, C., Gahinet, P., Chilali, M.: Multi-objective output-feedback control via LMI optimization. IEEE Trans. on Automatic Control 42, 896–911 (1997)
85. Scorletti, G., Ghaoui, L.E.: Improved linear matrix inequality conditions for gain scheduling. In: Proc. of the 34th IEEE Conf. on Decision and Control, pp. 3626–3631 (1995)
86. Shimomura, T., Fujii, T.: Multi objective control design via successive overbounding of quadratic terms. In: Proc. of the 39th Conf. on Decision and Control, pp. 2763–2768 (2000)
87. Spurgeon, S.K.: Hyperplane design techniques for discrete-time variable structure control systems. Int. J. of Control 55(2), 445–456 (1992)
88. Su, W.-C., Drakunov, S.V., Ozguner, U.: Constructing discontinuity surfaces for variable structure systems: A lyapunov approach. Automatica 32, 925–928 (1996)
89. Su, W.-C., Drakunov, S.V., Ozguner, U.: An $O(T^2)$ boundary layer in sliding mode for sampled-data systems. IEEE Trans. on Automatic Control 45(3), 482–485 (2000)
90. Takahashi, R.H.C., Peres, P.L.D.: H_2 guaranteed cost-switching surface design for sliding modes with nonmatching disturbances. IEEE Trans. on Automatic Control 44, 2214–2218 (1999)
91. Tang, C.Y., Misawa, E.A.: Discrete variable structure control for linear multivariable systems: the state feedback case. In: Proc. of American Control Conf., pp. 114–118 (1998)
92. Tang, C.Y., Misawa, E.A.: Discrete variable structure control for linear multivariable systems. Journal of Dynamic Systems, Measurement, and Control 122, 783–792 (2000)
93. Turner, M.C., Postlethwaite, I.: Nonlinear regulation in constrained input discrete-time linear systems. International Journal of Control 77(15), 1330–1342 (2004)
94. Turner, M.C., Postlethwaite, I., Walker, D.J.: Non-linear tracking control for multivariable constrained input linear systems. International Journal of Control 73(12), 1160–1172 (2000)
95. Umeno, T., Hori, Y.: Robust speed control of d.c servomotors using modern two degrees-of-freedom controller design. IEEE Transaction on Ind. Electronics 38(5), 363–368 (1991)
96. Utkin, V.I.: Variable structure system with sliding mode. IEEE Transaction on Automatic Control 22(2), 212–221 (1977)

97. Utkin, V.I.: Sliding Modes in Control Optimization. Springer, New York (1992)
98. Utkin, V.I.: In Variable structure and Lyapunov Control, ch. 5. Springer, U.K (1993)
99. Utkin, V.I., Guldner, J., Shi, J.: Sliding Mode Control in Electromechanical Systems. Taylor & Francis, Abington (1999)
100. Utkin, V.I., Shi, J.: Integral sliding mode in systems operating under uncertainty conditions. In: Proceedings of the 35th conference on decision and control, pp. 4591–4596 (1996)
101. Utkin, V.I., Yang, K.D.: Methods for constructing discontinuity planes in multidimensional variable structure systems. Automation and Remote Control 39, 1466–1470 (1979)
102. Wang, J., Brussel, H.V., Swevers, J.: Perfect tracking control with discrete sliding mode controller. Journal of Dynamic Systems, Measurement, and Control 125, 27–32 (2003)
103. Wang, Z., Goldsmith, P., Tan, D.: Improvement on robust control of uncertain systems with time-varying input delays. IET Control Theory and Applications 1(1), 189–194 (2007)
104. Xia, Y., Liu, G.P., Shi, P., Chen, J., Rees, D., Liang, J.: Sliding mode control of uncertain linear discrete-time systems with input delay. IET Control Theory Appl. 1(4), 1169–1175 (2007)
105. Xu, J., Pan, Y., Lee, T.H.: A gain scheduled sliding mode control scheme using filtering techniques with applications to multilink robotic manipulators. Journal of Dynamic Systems Measurement and Control Transactions of the ASME 122(4), 641–649 (2000)
106. Yan, M., Shi, Y.: Robust discrete-time sliding mode control for uncertain systems with time-varying state delay. IET Control Theory Appl. 2(8), 662–674 (2008)
107. Yasuda, K., Nakatsuji, Y.: Robust sliding mode control of uncertain systems. In: Proc. of the 1996 IEEE Workshop on Variable Structure Systems, pp. 15–19 (1996)
108. Yedavalli, R.K.: Robust root clustering for linear uncertain systems using generalized lyapunov theory. Automatica 29, 237–240 (1993)
109. Yilmaz, C., Hurmuzlu, Y.: Elimination of reaching phase from variable structure control. Journal of Dynamic Systems, Measurement and Control 122, 753–757 (2000)
110. Young, K.D., Utkin, V.I., Ozguner, U.: A control engineer's guide to sliding mode control. IEEE Trans.Control System Technology 7(3), 328–342 (1999)
111. Zhang, D.Q., Guo, G.X.: Discrete-time sliding mode proximate time optimal seek control of hard disc drives. IEE Proceedings 147(4), 440–446 (2000)

Index

Lecture Notes in Control and Information Sciences

Edited by M. Thoma, F. Allgöwer, M. Morari

Further volumes of this series can be found on our homepage:
springer.com

Vol. 392: Bandyopadhyay, B.; Deepak, F.; Kim, K.-S.:
Sliding Mode Control Using Novel Sliding Surfaces
137 p. [978-3-642-03447-3]

Vol. 391: Khaki-Sedigh, A.; Moaveni, B.:
Control Configuration Selection for Multivariable Plants
232 p. [978-3-642-03192-2]

Vol. 390: Chesi, G.; Garulli, A.; Tesi, A.; Vicino, A.:
Homogeneous Polynomial Forms for Robustness Analysis of Uncertain Systems
197 p. [978-1-84882-780-6]

Vol. 389: Bru, R.; Romero-Vivó, S. (Eds.):
Positive Systems
398 p. [978-3-642-02893-9]

Vol. 388: Jacques Loiseau, J.; Michiels, W.; Niculescu, S-I.; Sipahi, R. (Eds.):
Topics in Time Delay Systems
418 p. [978-3-642-02896-0]

Vol. 387: Xia, Y.; Fu, M.; Shi, P.:
Analysis and Synthesis of Dynamical Systems with Time-Delays
283 p. 2009 [978-3-642-02695-9]

Vol. 386: Huang, D.; Nguang, S.K.:
Robust Control for Uncertain Networked Control Systems with Random Delays
159 p. 2009 [978-1-84882-677-9]

Vol. 385: Jungers, R.:
The Joint Spectral Radius
144 p. 2009 [978-3-540-95979-3]

Vol. 384: Magni, L.; Raimondo, D.M.; Allgöwer, F. (Eds.):
Nonlinear Model Predictive Control
572 p. 2009 [978-3-642-01093-4]

Vol. 383: Sobhani-Tehrani E.; Khorasani K.;
Fault Diagnosis of Nonlinear Systems Using a Hybrid Approach
360 p. 2009 [978-0-387-92906-4]

Vol. 382: Bartoszewicz A.; Nowacka-Leverton A.;
Time-Varying Sliding Modes for Second and Third Order Systems
192 p. 2009 [978-3-540-92216-2]

Vol. 381: Hirsch M.J.; Commander C.W.; Pardalos P.M.; Murphey R. (Eds.)
Optimization and Cooperative Control Strategies: Proceedings of the 8th International Conference on Cooperative Control and Optimization
459 p. 2009 [978-3-540-88062-2]

Vol. 380: Basin M.
New Trends in Optimal Filtering and Control for Polynomial and Time-Delay Systems
206 p. 2008 [978-3-540-70802-5]

Vol. 379: Mellodge P.; Kachroo P.;
Model Abstraction in Dynamical Systems: Application to Mobile Robot Control
116 p. 2008 [978-3-540-70792-9]

Vol. 378: Femat R.; Solis-Perales G.;
Robust Synchronization of Chaotic Systems Via Feedback
199 p. 2008 [978-3-540-69306-2]

Vol. 377: Patan K.
Artificial Neural Networks for the Modelling and Fault Diagnosis of Technical Processes
206 p. 2008 [978-3-540-79871-2]

Vol. 376: Hasegawa Y.
Approximate and Noisy Realization of Discrete-Time Dynamical Systems
245 p. 2008 [978-3-540-79433-2]

Vol. 375: Bartolini G.; Fridman L.; Pisano A.; Usai E. (Eds.)
Modern Sliding Mode Control Theory
465 p. 2008 [978-3-540-79015-0]

Vol. 374: Huang B.; Kadali R.
Dynamic Modeling, Predictive Control and Performance Monitoring
240 p. 2008 [978-1-84800-232-6]

Vol. 373: Wang Q.-G.; Ye Z.; Cai W.-J.; Hang C.-C.
PID Control for Multivariable Processes
264 p. 2008 [978-3-540-78481-4]

Vol. 372: Zhou J.; Wen C.
Adaptive Backstepping Control of Uncertain Systems
241 p. 2008 [978-3-540-77806-6]

Vol. 371: Blondel V.D.; Boyd S.P.; Kimura H. (Eds.)
Recent Advances in Learning and Control
279 p. 2008 [978-1-84800-154-1]

Vol. 370: Lee S.; Suh I.H.; Kim M.S. (Eds.)
Recent Progress in Robotics: Viable Robotic Service to Human
410 p. 2008 [978-3-540-76728-2]

Vol. 369: Hirsch M.J.; Pardalos P.M.; Murphey R.; Grundel D.
Advances in Cooperative Control and Optimization
423 p. 2007 [978-3-540-74354-5]

Vol. 368: Chee F.; Fernando T.
Closed-Loop Control of Blood Glucose
157 p. 2007 [978-3-540-74030-8]

Vol. 367: Turner M.C.; Bates D.G. (Eds.)
Mathematical Methods for Robust and Nonlinear Control
444 p. 2007 [978-1-84800-024-7]

Vol. 366: Bullo F.; Fujimoto K. (Eds.)
Lagrangian and Hamiltonian Methods for Nonlinear Control 2006
398 p. 2007 [978-3-540-73889-3]

Vol. 365: Bates D.; Hagström M. (Eds.)
Nonlinear Analysis and Synthesis Techniques for Aircraft Control
360 p. 2007 [978-3-540-73718-6]

Vol. 364: Chiuso A.; Ferrante A.; Pinzoni S. (Eds.)
Modeling, Estimation and Control
356 p. 2007 [978-3-540-73569-4]

Vol. 363: Besançon G. (Ed.)
Nonlinear Observers and Applications
224 p. 2007 [978-3-540-73502-1]

Vol. 362: Tarn T.-J.; Chen S.-B.; Zhou C. (Eds.)
Robotic Welding, Intelligence and Automation
562 p. 2007 [978-3-540-73373-7]

Vol. 361: Méndez-Acosta H.O.; Femat R.; González-Álvarez V. (Eds.):
Selected Topics in Dynamics and Control of Chemical and Biological Processes
320 p. 2007 [978-3-540-73187-0]

Vol. 360: Kozlowski K. (Ed.)
Robot Motion and Control 2007
452 p. 2007 [978-1-84628-973-6]

Vol. 359: Christophersen F.J.
Optimal Control of Constrained Piecewise Affine Systems
190 p. 2007 [978-3-540-72700-2]

Vol. 358: Findeisen R.; Allgöwer F.; Biegler L.T. (Eds.): Assessment and Future Directions of Nonlinear Model Predictive Control
642 p. 2007 [978-3-540-72698-2]

Vol. 357: Queinnec I.; Tarbouriech S.; Garcia G.; Niculescu S.-I. (Eds.):
Biology and Control Theory: Current Challenges
589 p. 2007 [978-3-540-71987-8]

Vol. 356: Karatkevich A.:
Dynamic Analysis of Petri Net-Based Discrete Systems
166 p. 2007 [978-3-540-71464-4]

Vol. 355: Zhang H.; Xie L.:
Control and Estimation of Systems with Input/Output Delays
213 p. 2007 [978-3-540-71118-6]

Vol. 354: Witczak M.:
Modelling and Estimation Strategies for Fault Diagnosis of Non-Linear Systems
215 p. 2007 [978-3-540-71114-8]

Vol. 353: Bonivento C.; Isidori A.; Marconi L.; Rossi C. (Eds.)
Advances in Control Theory and Applications
305 p. 2007 [978-3-540-70700-4]

Vol. 352: Chiasson, J.; Loiseau, J.J. (Eds.)
Applications of Time Delay Systems
358 p. 2007 [978-3-540-49555-0]

Vol. 351: Lin, C.; Wang, Q.-G.; Lee, T.H., He, Y.
LMI Approach to Analysis and Control of Takagi-Sugeno Fuzzy Systems with Time Delay
204 p. 2007 [978-3-540-49552-9]

Vol. 350: Bandyopadhyay, B.; Manjunath, T.C.; Umapathy, M.
Modeling, Control and Implementation of Smart Structures 250 p. 2007 [978-3-540-48393-9]

Vol. 349: Rogers, E.T.A.; Galkowski, K.; Owens, D.H.
Control Systems Theory and Applications for Linear Repetitive Processes
482 p. 2007 [978-3-540-42663-9]

Vol. 347: Assawinchaichote, W.; Nguang, K.S.; Shi P.
Fuzzy Control and Filter Design for Uncertain Fuzzy Systems
188 p. 2006 [978-3-540-37011-6]

Vol. 346: Tarbouriech, S.; Garcia, G.; Glattfelder, A.H. (Eds.)
Advanced Strategies in Control Systems with Input and Output Constraints
480 p. 2006 [978-3-540-37009-3]

GPSR Compliance

The European Union's (EU) General Product Safety Regulation (GPSR) is a set of rules that requires consumer products to be safe and our obligations to ensure this.

If you have any concerns about our products, you can contact us on ProductSafety@springernature.com

In case Publisher is established outside the EU, the EU authorized representative is:

Springer Nature Customer Service Center GmbH
Europaplatz 3
69115 Heidelberg, Germany

Batch number: 10682617

Printed by Printforce, the Netherlands